Protocol Specification and Testing

Protocol Specification and Testing

K. Tarnay

Central Research Institute for Physics
Budapest, Hungary

PLENUM PRESS · NEW YORK AND LONDON

ISBN-13: 978-1-4684-5780-3 e-ISBN-13: 978-1-4684-5778-0
DOI: 10.1007/ 978-1-4684-5778-0

Library of Congress Catalog Number 90-053587

Joint edition published by Plenum Press, New York
A Division of Plenum Publishing Corporation, New York, U.S.A.
and
Akadémiai Kiadó, Budapest, Hungary

Softcover reprint of the hardcover 1st edition 1991

CONTENTS

PREFACE XV
ACKNOWLEDGEMENTS XVII
INTRODUCTION XIX

Computer communication
Communication protocols
Formal description techniques
Applications
Outline of the book

PART I

COMMUNICATION PROTOCOLS OVERVIEW

1. THE REFERENCE MODEL AND ITS LAYERS

1.1 Computer Networks, Open Systems 3

The concept of computer networks
The origins of computer networks
The operation of computer networks
The beginnings of the Reference Model

1.2 The Reference Model 11

The essence of the Reference Model
The elements of the Reference Model
Functional division of the distributed system
Management in open system interconnection

1.3 Layers, Layer Terms and Layer Elements 14

The general principles of layering
Layer elements
Identifiers. Naming and addressing

1.4 Layer Functions 18

The general layer functions
Connection establishment and release
Data transfer. Data units
Error handling

1.5 Layer Conventions 22

The model of the layer services
Service primitives
Time sequence diagrams

1.6 Protocols 25

The basic communication concept of Open Systems: the protocol
Protocol features and their control
The protocol and its environment

2. THE SEVEN LAYERS

2.1 About the Seven Layers 29

The seven layers
The functions of the seven layers

2.2 The Physical Layer and Its Protocols 32

The physical layer and its services
The data transfer
Mechanical, electrical, procedural and functional characteristics of the physical layer

2.3 The Data Link Layer and Its Protocols 36

The data link layer and its services
Data link functions
Classification of the data link protocols

2.4 Binary Synchronous Communication (BSC) Protocol 39

The main feature of BSC
The character set of BSC
Time-out

2.5 High-level Data Link Control (HDLC) Protocol 43

The fundamental concept of HDLC
Frame types
Error-free operation of the HDLC
Erroneous operation of the HDLC

2.6 The Network Layer and Its Protocols 53

The network layer services
X.25 packet level procedure
Packet formats
Routing

2.7 The Transport Layer and Its Protocols 59

The transport layer and the transport services
Transport protocol classes

2.8 The Session Layer and Its Protocols 66

The session layer
Session layer functions
Operation of the session layer

2.9 The Presentation Layer and Its Protocols 68

The presentation layer and its services
Transfer syntax and presentation context

2.10 The Application Layer and Its Protocols 70

The application layer services and functions
The application process
General application layer protocols
Application layer structure

3. ABSTRACT MODEL OF TESTING

3.1 Purpose of Testing 77

Specification, verification and testing
Conformance testing
Performance measurement

3.2 Network Environment 79

Characteristics of the network
Queues in the network
The traffic

3.3 Error Generators 82

Relationship between the network topology and the errors
Relationship between the layered structure and the errors
Error generators of the protocol testers

3.4 Conformance Testing 89

Requirements
The classification of conformance testing

3.5 The Abstract Model of Conformance Testing 93

The fundamental concept of abstract testing
Methods of conformance testing
The general logical architecture of conformance testers
The physical architecture of conformance testers
An example on conformance testing

3.6 Performance Testing 99

Requirements
The schedule of performance testing
Parameters of performance testing in the OSI model

PART II

FORMAL DESCRIPTION TECHNIQUES

4. STATE-TRANSITION BASED MODELS

4.1 Overview of Finite-state Machines 107

Basic set theory
Computability. Turing machines
Finite-state machines
State-transition tables and graphs

4.2 Examination of Finite-state Machines 114

The properties of finite-state machines
Equivalence of finite-state machines
State identification with successor tree

5. GRAPH MODELS

5.1 Petri Nets 123

Fundamental concepts of Petri nets
Petri net features
The subgraphs of the Petri net
Mathematical description of Petri nets

5.2 Modified Petri Nets 132

Classification of modified Petri nets
Time Petri nets
E-nets
Numerical Petri nets

5.3 Data Flow Graphs 142

Graph model of behaviour
Data flow graph models

5.4 **Examples of Graph Models** 146

Examples of Petri nets
Graph models of protocol components

6. ALGEBRAS AND FORMAL LANGUAGES

6.1 **Calculus of Communicating Systems** 151

Introduction
Basic ideas
The syntax and semantics of CCS
Equivalences
Example

6.2 **A Specification Language for Abstract Data Types** 160

Introduction
Abstract data types
Language concepts of ACT ONE
Abstract data types of ACT ONE
Atomic specification
Complex specification

6.3 **Formal Languages** 169

Introduction to formal languages
Classification of grammars

7. SPECIFICATION LANGUAGES FOR COMMUNICATION

7.1 **Specification and Description Language (SDL)** 175

Overview of SDL
General definition of SDL
Basic concepts
SDL process diagram
From the basic concepts to the structuring concepts
Partitioning
Refinement
Data concepts in SDL
Operators
Axioms
Inheritance
Tools for SDL

7.2 Architectural Specification Concept 197

7.3 ESTELLE, an Extended Finite-state Machine Specification Language . . . 200

Overview of ESTELLE
Transitions
Modules
Interaction points and channel type
Specification
Example
Software tools for ESTELLE

7.4 LOTOS, a Language of Temporal Ordering Specification 213

Overview of LOTOS
Interactions between processes
Behaviour expressions
LOTOS data types
Recursion and nondeterminism

PART III

APPLICATIONS

8. ALTERNATING-BIT PROTOCOL SPECIFICATION

8.1 State-transition Graph of the Alternating-bit Protocol 229

8.2 Alternating-bit Protocol Description with Petri Net 234

A Petri net specification complemented with variables

8.3 ESTELLE Specification of the Alternating-bit Protocol 240

The modules
Declaration of the data transfer submodule
Transitions

9. PROTOCOL SPECIFICATIONS IN THE DATA LINK LAYER

9.1 The Hybrid HDLC Model 247

9.2 A Formal Grammar Description of HDLC 252

Essence of the method
Half-duplex models
Improved model with sequence numbering
Full-duplex models
Conclusions

9.3 Formal Description of LAPB with SDL/GR 263

The structure of LAPB description
The processes
Signals
The data of the SDL processes
The process diagrams: the SDL-graphs

10. FORMAL DESCRIPTION OF TRANSPORT PROTOCOLS

10.1 Transport Protocol Description with Finite-state Machine 281

10.2 Transport Protocol Description with SDL-graph 284

Extended state-transition graph of the transport protocol
SDL/GR description of the transport protocol

10.3 Numerical Petri Nets of the Transport Protocol 288

Transitions
Enabling conditions for input arcs
Firing rules
Conclusions

10.4 Data Flow Graph for Transport Protocol Description 293

10.5 LOTOS Specification of the Transport Protocol 295

Structure of the specification
Data type definitions
Specification of the transport connection in LOTOS

11. CONFORMANCE TESTERS

11.1 Initial Protocol Testing 305

11.2 Conformance Test Centre in the National Bureau of Standards 306

Requirements
The test system
Test strategy

11.3 A Conformance Tester 310

Objectives
Manual conformance tester
State-transition table controlled conformance tester

11.4 Testing and Diagnosis Aids for Higher Level Protocols 312

11.5 AT&T Protocol Test Centre 313

Requirements
Test centre architecture
Test sequences

12. TEST SEQUENCES

12.1 Test Sequence Generation 321

Transition tour method

12.2 W-method 324

12.3 Gönenc Method 326

12.4 Testing the NATHAN Application Protocol 331

12.5 Automatic Test Sequence Generation 340

Automatic testing
The model used
The tools

GLOSSARY 343

REFERENCES 347

STANDARDS 357

ABBREVIATIONS 359

INDEX 363

P R E F A C E

The increasing number of computer networks has aroused users' interest in many and various fields of applications, in how a computer network can be built, and in how it may be used. The fundamental rules of computer networks are the protocols. "A protocol is a set of rules that governs the operation of functional units to achieve communication" [STA-86]. The book follows a practical approach to protocol specification and testing, but at the same time it introduces clearly and precisely the relevant theoretical fundamentals.

The principal objectives of this work are:

- *to familiarize readers with communication protocols,*
- *to present the main, formal description techniques,*
- *to apply various formal description techniques to protocol specification and testing.*

It is considered that the readership will primarily consist of protocol developers, protocol users, and all who utilize protocol testers. Secondly the book is suggested for postgraduate courses or other university courses dealing with communication networks and data communication. A large part of the book provides a comprehensive overview for managers; some parts are of especial interest to postal organizations.

The book consists of three parts: the first part introduces the OSI Reference Model, it provides an overview of the most frequently used protocols and explains the fundamentals of protocol testing. The second part familiarizes readers with the methods used for protocol specification, generation, and testing. Finite-state machines, formal grammars, Petri nets and some specification languages (SDL, ESTELLE, LOTOS) are discussed in a pragmatic style. The third part deals with applications. It contains many practical examples based on real-world protocols. The protocols introduced in the first part are specified using the methods of the second part. Finally some test methods based on the specifications are analysed and some characteristic conformance testers are discussed.

Most of the chapters contain some examples. The most important international data communication standards, literature and abbreviations are summarized at the end of the book.

A brief glossary summarizes the essential terms and expressions.

Occasionally the book is repetitive. The reason for this is that the first explanation offers a preliminary approach and the explanations become deeper and deeper, step-by-step.

Apart from its use on university and other courses, the book is also suitable for self-study.

ACKNOWLEDGEMENTS

In that this book was prepared in many stages over a long period of time, I have been fortunate in having been able to consult many people. First of all I am grateful for the constructive remarks of László Csaba, Csaba Szabó and Harvey Shenker who reviewed this work in manuscript form.

The Central Research Institute for Physics, Budapest supported my work. This support was provided by János Szlankó, our director and József Lukács, our scientific director.

Professor Gregor von Bochmann invited me for a semester (1986) to the University of Montreal as a guest professor, I should like to thank him for having given me the opportunity to participate in many fruitful discussions. Professor Jan Gecsei read the manuscript and suggested many improvements, I am especially grateful for his critical remarks and useful ideas.

Those parts of the book on conformance testing are based on my practical work for S.W.I.F.T. (Society for Worldwide Interbank Financial Telecommunication S.C.) in Brussels from 1983-85, during which time I was able to discuss the main questions with Joseph Bremer and Michel Paris.

Dave Rayner, the Rapporteur of the OSI Conformance Testing Group (ISO TC 97/SC21/WG1) created a highly congenial atmosphere at the various meetings and this led to many results; as the Hungarian expert of the group I have been able to apply many of the resulting ideas in my book.

From 1983 up till the present, I have been delivering postgradute courses at the Technical University of Budapest about protocol specification and testing, I am particularly grateful to my students for the fruitful discussions.

Various parts of the book were discussed with József Harangozó, Behcet Sarikaya, Hartmut König, Péter Ecsedy Tóth, Péter Hunya and Pál Tőke, I should like to express my thanks to them for their contributions.

I am also indebted to Ramesh Ahooja, Denise Ambrózy, Mihály Bohus, László Böszörményi, Rachida Dssouli, András Ercsényi, Reinhard Gotzhein, Imre Kovács, László Kovács, Sándor Mazgon, Abdulmagied Sdaa, István Sarkadi Nagy, Tibor Szentiványi, Ferenc Telbisz, Sarolta Wagner-Dibuz and Jin Rong Zhao.

I am grateful to László Zsolt Varga for the correction of the camera-ready manuscript.

The manuscript was painstakingly typed by Éva Végvári; the really handsome figures are the work of Judit Kacsó.

To my husband and son without whose encouragement and patience this work would never have been completed.

INTRODUCTION

Computer Communication

The world is moving towards what we might call the information society. Ideally, everybody in this information society should have an up-to-date computing environment with an easy-to-achieve communication background. The principal means of integrating the computing environment and the communication tools is computer communication. Computer communication is responsible for the data message exchange between the computerized source station and the destination station. The stations are connected in a network.

As a rough guide, let us consider a computer network as being composed of a set of computer stations. The user of a computer network works on a host computer, which may be a desktop workstation or a time-shared computer. The host computer is connected, generally, through a local area network to a wide area network. The services provided by the computer network can be of great benefit to engineers, managers and computists. It is to be hoped that the use of these services will lead to increased productivity.

The network services spread from file transfer and database access to the most commonly available electronic mail. A number of new services are due to appear in the next few years realizing multi-media data transfer, where voice, text, and graphics can be intermixed in a single message. The benefits of computer networks include increased opportunity for collaboration with colleagues, rapid dissemination of information and a real revolution in industry.

The book comprises three parts: setting out from the fundamentals of computer communications, the protocols and their mathematical descriptions are discussed, finally the mathematical models are utilized for protocol specification and testing.

The author of the book hopes that she is not being too ambitious in endeavouring to provide a unified view of these three topics.

Communication Protocols

The various types of computer systems connected in a computer network are unable to "talk" to each other because, having been created by different manufacturers, they do not have the same native language. It was therefore necessary to develop sequences of messages with specific formats and meanings; these are known as protocols. The basic protocol functions are addressing, error control, flow control and synchronization.

The basic functional decomposition of the network is related to the layers and each layer has its own protocol or protocols. This layered structure will be discussed in Part I. One of the first protocols was the BSC suggested and applied by IBM. The activities of IBM and other manufacturers gave the impetus to the various organizations for standardization (e.g. ISO, CCITT) to work on interconnecting the systems of different vendors.

It is by no means easy to find a mathematical model which provides an unambiguous description of the protocols. The basic ideas are discussed in the next section.

Formal Description Techniques

Formal description techniques are based on formal models. Formal models (e.g. the finite-state automata, the abstract state machines) describe the general system behaviour. These general models can be used for protocol specification and there are protocol-specific formal models, too.

The following are general formal models:

— transition models
  • finite-state automaton,
  • abstract state machine,
  • Petri net,
  • grammars,
— programming languages
  • Pascal, C, ADA, MODULA,
  • PROLOG,
— algebras
  • process algebra (e.g. CCS),
  • type algebra (e.g. ACT ONE).

The protocol-specific formal models were developed in ISO and CCITT for the service definitions and protocol specifications. The ISO has two languages:

— ESTELLE, based on state transitions
— LOTOS for temporal ordering specification.

ESTELLE is a hybrid model because the transitions are extended by programming language variables. LOTOS has a data type definition part and a process algebra part. Both have character-oriented notations.

CCITT has introduced a Specification and Description Language (SDL) with a high illustrative power. The hierarchical structure and logical understanding are well supported by the graphical notations.

The formal description techniques are discussed in Part II and the graphical representations are preferred in this book.

The next section summarizes how these formal descriptions can be used in the protocol field.

Applications

The application of the formal description technique is becoming increasingly important [BOCH-87]. Reliable and quick specification requires easy-to-use methods and tools. The formal methods allow a more systematic approach to protocol validation, implementation and testing, compared with the traditional use of protocol specifications given in a natural language.

The different development activities are related to specification, implementation, and testing. The specification development has two phases:

- specification creation
  this is the representation of the specification in a formal way,
- specification validation
  this is the control of consistency of the specification created with the specification development.

The implementation development includes the determination of the communication protocols and services, the creation of the implementation code, and the validation of the code.

A particular case of validation is conformance testing where the emphasis is on the validation of consistency of the implementation with the specification of the protocol to be implemented.

Great efforts are being made to produce automated tools for protocol development, they can improve the efficiency and the time requirements significantly.

The third part of the book discusses some protocol specifications and explains the basic ideas and tools for conformance testing.

Outline of the Book

This chapter is an introduction to the entire book. A brief summary of the remaining chapters follows.

The Reference Model and Its Layers

This introductory section shows the evolution from initial computer networks to the layered approach realized as an open systems interconnection (OSI) model and provides an overview of the layer terms, functions and conventions. The protocols, as the basic communication concept of the open systems, are presented.

The Seven Layers

Chapter 2 describes the functions of the seven layers, and the main protocols in each layer are discussed.

Abstract Models of Testing

The links between testing and specification and verification are examined and the fundamental terms of conformance testing are explored.

State Transition Based Models

Chapter 4 summarizes the main ideas of finite-state machines necessary for protocol specification and conformance testing.

Graph Models

Chapter 5 provides an overview of the various graph models used in protocol engineering. Graph models from various Petri nets to data flow graphs are examined.

Algebras and Formal Languages

The Calculus of Communicating Systems (CCS) and ACT ONE are the components of the specification language LOTOS. Chapter 6 explains how the interactions and concurrency can be described by CCS and how the abstract data type definition is used in ACT ONE. This section offers an introduction to formal grammars as a means of understanding protocol specification with formal grammars.

Specification Languages for Communication

Chapter 7 discusses the SDL, ESTELLE and LOTOS languages.

Alternating Bit Protocol Specification

The first attempts to specify a protocol dealt with alternating bit protocols. This chapter describes three solutions, viz. state transition graph, Petri net and ESTELLE.

Protocol Specification in the Data Link Layer

Three models (a hybrid model, a formal grammar description, and an SDL graphical version) are examined in Chapter 9.

Formal Description of Transport Protocol

The transport protocol is the keystone of the OSI architecture, therefore five formal models are discussed: finite-state machine, SDL graph, numerical Petri net, data flow graph and LOTOS.

Conformance Testers

Chapter 11 provides an overview of the evolution of conformance testers — from initial manual testers to automated conformance test centres.

Test Sequences

The conformance test methods are based on test sequence generation. The main problems are discussed and illustrated with examples.

PART I
COMMUNICATION PROTOCOLS OVERVIEW

1. THE REFERENCE MODEL AND ITS LAYERS

1.1 COMPUTER NETWORKS, OPEN SYSTEMS

The Concept of Computer Networks

A net is a set of nodes connected by arcs. The computer network is a net, too, but its nodes are data (information) processing systems and its arcs are data communication channels. In other words, the computer network is a system of communicating computers automatically exchanging information among the machines and interactively sharing the resources available. Interactive sharing means that any node can access any other remote node of the network as if it was a local resource. (The expression "network" is generally used for a computer network. If not, the readers' attention will be drawn to it.) The resources can be either hardware or software resources. The sharing of resources is economically effective; a single, very special, rarely used resource can become common property for all network users. The advantage is not only economic, but the power of the computers is multiplied too. Dynamic file access, the sharing on remote data and remote programs is not only useful but it also allows the homogeneous loading of the network.

A user may wish to become a member of a network in order to obtain the required services quickly, reliably and at a reasonable price. It is also advantageous in that the user can learn the usage of the system easily.

An appropriate question is with whom it is worthwhile to establish a network and to maintain it. It is primarily for big enterprises, government offices, and multinational firms whose plant and administrative centres are widely dispersed and who are in need of different computer services at many places and where information exchange between the remote nodes is permanently required. Computer networks were originally developed for the reservation of seats in aircraft, for financial transactions among banks, for process control and for office systems. Current local area networks will not be discussed. Local area networks use the general principles of computer networks, but the distances are restricted.

Computer networks can be classified from several different points of view:

- application (place reservation, bank transfer, process control, office automation, computerized manufacturing, etc.),
- topology (star, ring, distributed, chain, tree, etc.),
- the number of nodes,
- the similarity or dissimilarity of nodes (homogeneous or heterogeneous networks),
- according to the type of node (network with microcomputers, minicomputers, or large computers),

— the type of channel (cable, radio communication),
— who is the owner of the network (the Postal Administration, a large enterprise, etc.),
— geographical distribution (local, national, multinational).

IBM carried out the first experiments in data communication in 1940. At the request of the US Air Force a system was started to enable a phone link to be established between a card-punch and a card-reader.

In the same year the first attempt was made to operate an interactive terminal. At a meeting of the American Mathematical Society a telex terminal (an electrical typewriter) was connected through a telegraph network to a computer built up from relays and operating in the Bell Laboratory about 100 km from the meeting.

There were two basic causes underlying the origin of computer networks: one was based on the need to operate remote terminals, the spread of interactive operation and the development of telecommunication; the second was to allow the greater utilization of computers.

The Origins of Computer Networks

The first steps towards real computer networks were made at the beginning of the sixties. More input-output devices were directly connected to the computer in order to utilize it more efficiently (Fig. 1.1). The direct connection was very simple, but also caused many problems:

— the speed of the information flow between the devices and the computer memory is different,
— the information unit of the I/O devices is different from a word of memory,
— more than one device want to communicate with the computer memory.

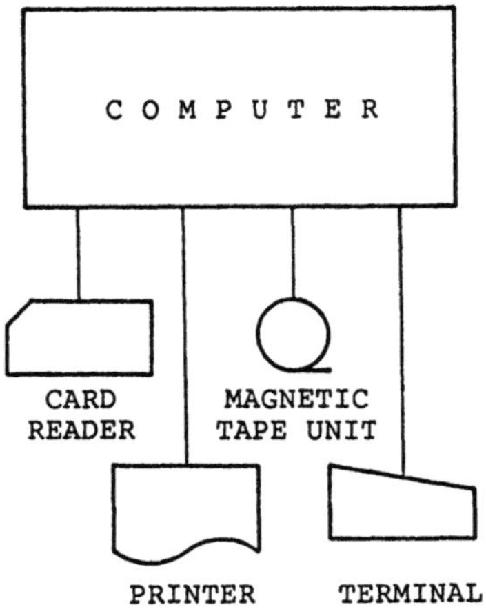

Fig. 1.1. I/O devices directly coupled to the computer

In order to solve these problems the channel principle was worked out. One channel can control many devices and can convert the unit of incoming information as well as its speed and its format according to the requirements of the computer memory. It can also prevent conflicts between messages coming from different terminals (Fig.1.2).

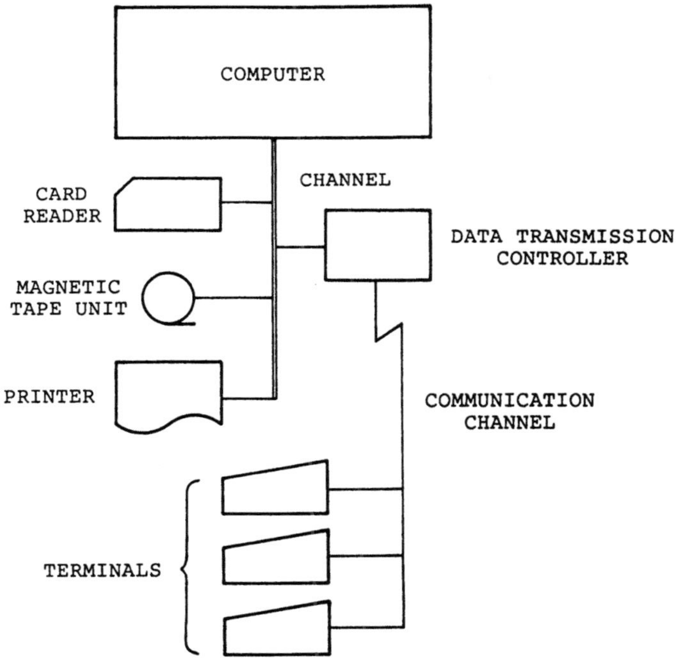

Fig. 1.2. I/O devices coupled to the computer through a channel

The next step was the writing and assembling of macro instructions to help with the handling of several terminals. The collections of macro instructions comprise the various telecommunication access methods. A typical example is the BTAM (Basic Telecommunication Access Method) developed by IBM for handling terminals, using character-oriented procedures. The essential components of typical (remote) data-communication systems are the different telecommunication access methods (Fig. 1.3). It is by no means easy to use these methods, because one needs to know all the features of the data communication network and all the characteristics of the terminals.

All access methods deal with link access control and terminal handling, just as with character code conversion. If the IBM 2740 typewriter-based terminals use different character codes (e.g. EBCDIC or ASCII), then the access method should include different programs to control the different terminals.

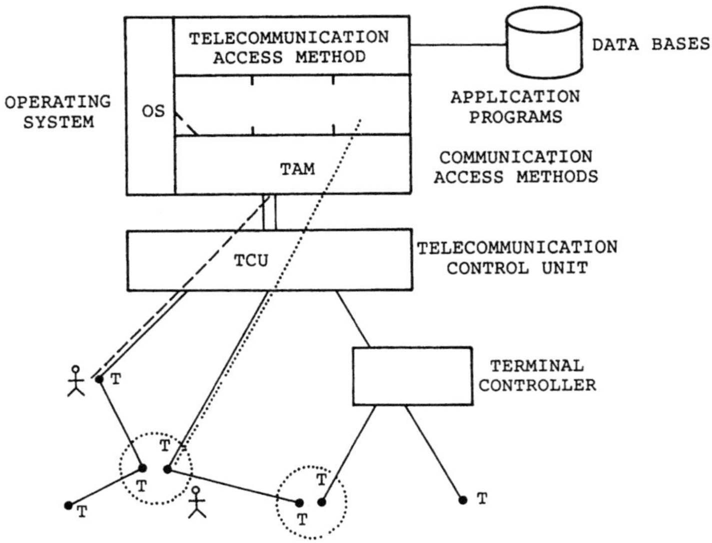

Fig. 1.3. Initial telecommunication system

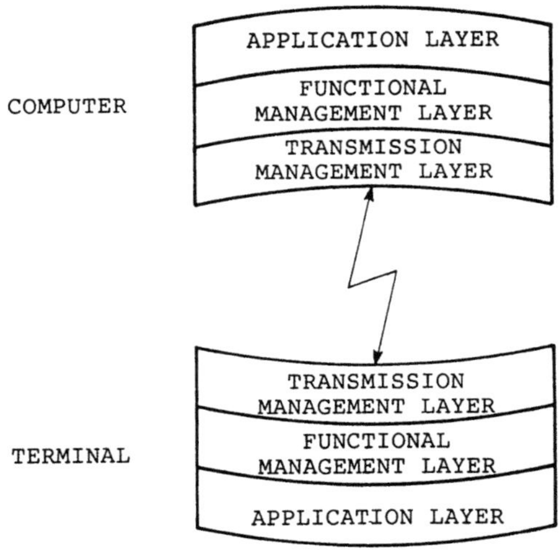

Fig. 1.4. Origin of the layered structure

IBM initiated the layered structure to solve the problems caused by the various telecommunication access methods. The nodes were decomposed into functional blocks:

— transmission management,

— function management,

— application.

Each functional block creates a layer. The layered structure according to the IBM partitioning can be seen in Fig. 1.4. The introduction of the layered structure was a milestone in the evolution of computer communication networks. Very soon other large computer firms (DEC, Burroughs, Honeywell, etc.) also adopted the layered structure and this became the basis of the international network standards (ISO, CCITT, etc.) [DAV-BAR-79].

The Operation of Computer Networks

The object of a computer network operation is to create a full access path between the various end-users. This access path includes the physical connection, but also the functions and actions allowing the connection of two end-users despite transmission errors, different rates of information flow and various code formats (Fig. 1.5).

What kind of requirements should be fulfilled to allow a full access path?

Figure 1.6 shows the functional grouping that was first proposed by IBM.

When the network is connected through a long-distance telephone connection the transmission path means a carrier channel. This is why an analog transmission is established and modems convert the information given in bits into an analog signal and will then convert it back (modulation and demodulation). The arriving bit sequence must be checked to make sure that it is an error-free copy of the bit sequence sent. This checking is one of the tasks of the data link protocols (also called the "data link control procedures"). The information transport is carried out "in pieces", e.g. by frames or blocks. Redundant bits are added to each frame and these indicate transmission errors at reception. The sender repeats the information transfer after an erroneous frame.

The effectiveness of the transmission is increased by good utilization of the interim nodes and by the different methods of establishing the connection.

The connection will remain by circuit-switching as long as the user does not release and disconnect it. Packet-switching allows the sharing of the interim resources in an equal proportion. The user's messages are broken up into packets or frames, suitably addressed, and the forwarding of packets or frames is from node to node and independent of the other frames or packets. This is advantageous because the packets passing through the nodes are shorter so their repetition causes less additional load in cases of error. The buffer size of the nodes is determined by the packet length. A node preserves (stores) the packet to be sent until the next node sends an acknowledgement. Therefore packet-switching is also called the store-and-forward principle. Every packet contains a source address and a next node address to determine the route.

The task of routing is to define the route. The routing should consider the failure of a network section due to a line error or node error and it should generate an alternative route. Adaptive routing considers the load of different network sections and it tries to ensure an equal load in the network with the help of the route selection.

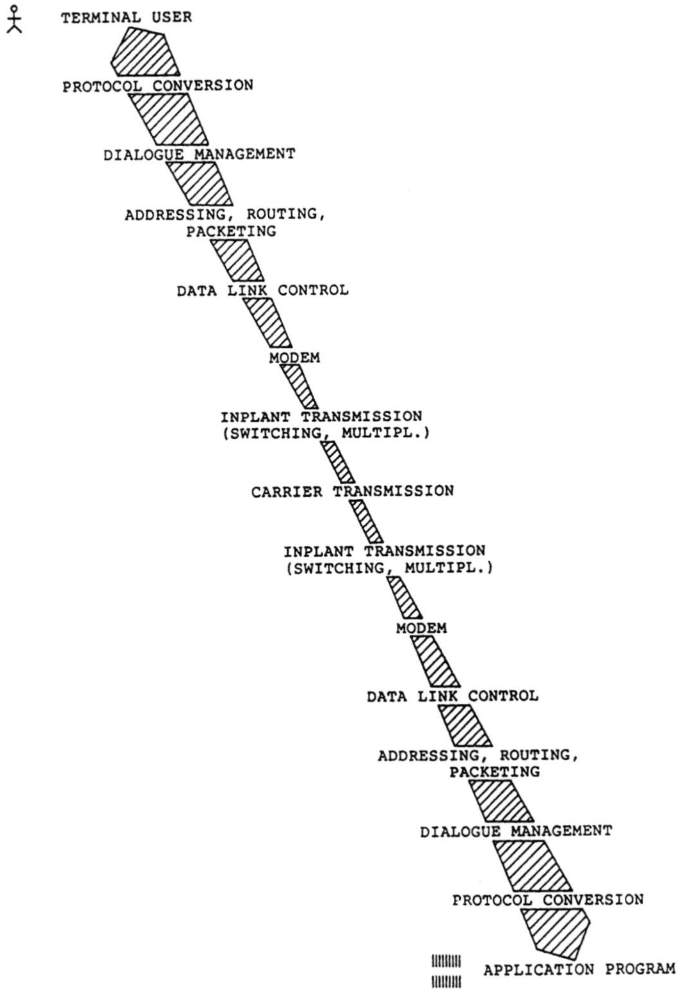

Fig. 1.5. Principle of the computer network operation

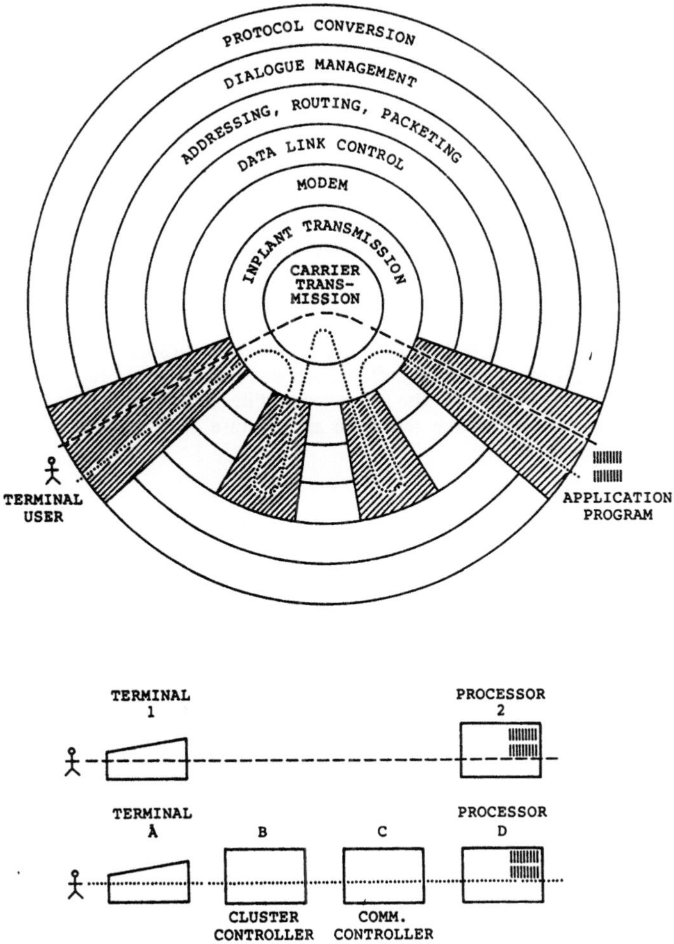

Fig. 1.6. Grouped network functions

The flow control mechanism is responsible for the control of the whole information flow in the network. The flow control determines either the number of all message units in the network or the maximum number of message units and when this maximum is exceeded the entry of further messages is restricted.

The computer network is responsible for the assembling of the disassembled packets at the receiving end and for the different code and format conversions.

The Beginnings of the Reference Model

Computer networks are either homogeneous distributed systems or heterogeneous ones. In homogeneous distributed systems the connected computer system elements belong to the same computer family. Their connection is simple and numerous technical questions never arise. Homogeneity is still not a general solution because computers with different features are much more suitable for the user's requirements and computer firms have no wish to lose the unique characteristics of their own products. Therefore the concept of a reference architecture was introduced to facilitate the connection of heterogeneous systems. The reference architecture constitutes an unambiguous definition. The essence of the reference architecture, is its openness that allows connection and expansion (Fig.1.7).

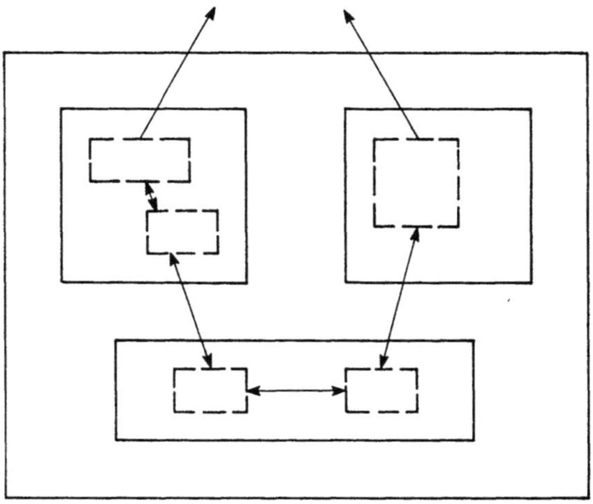

Fig. 1.7. Reference architecture

The reference architecture connects heterogeneous computer systems through reference interfaces. Reference interfaces are interfaces that provide a connection to the reference architecture. Everything inside the system can be seen only through the interface. The Open System Interconnection (OSI) is based on this principle.

The open system can be connected to other systems that have the standard requirements of the Reference Model. The parts of the system are:

— one or more computers,
— the software, the peripherials, the terminals, operators, physical processes belonging to the computers and the tools for information exchange [BOH-84], [CSA-81], [CSA-86], [SEB-87], [STA-85], [TAR-76], [TAR-83].

1.2 THE REFERENCE MODEL

The Essence of the Reference Model

The behaviour of a reference system is modelled by a Reference Model (RM). The Reference Model is an abstract model. The special modelling technique of the Reference Model means that every real open system is described by an abstract open system built up from the layers of abstract subsystems. The abstract open system defines the behaviour of a real open system as a whole and has no preconditions for its internal organization. Hence the Reference Model identifies the main functional blocks (the layers), and it defines the relationships between the blocks.

The Reference Model is an abstract model built up from communicating systems. At the end of the seventies the International Standards Organisation (ISO) began developing the international standard of the Reference Model for connecting open systems. The object of this standard is to create a common base to allow the coordination of the development of all standards that deal with the interconnection of systems. This international standard unambiguously defines concepts and ensures a functional framework for the various international groups of experts that are working on the Open Systems Interconnection Architecture (OSIA) and it allows them to develop the individual parts of the architecture independently of each other.

The Reference Model is a connection-oriented abstract model because the systems are considered on the basis of their communications. The Reference Model helps the implementation only indirectly by giving the appropriate principles. The Reference Model is a measure and a basis for comparison that supports the unified interpretation of the functional decomposition and the open systems.

The Elements of the Reference Model

The architecture of the Reference Model can be seen in Fig. 1.8.

The elements of the Reference Model are

— the system,
— the application entity,
— the connection,
— the physical interconnection media.

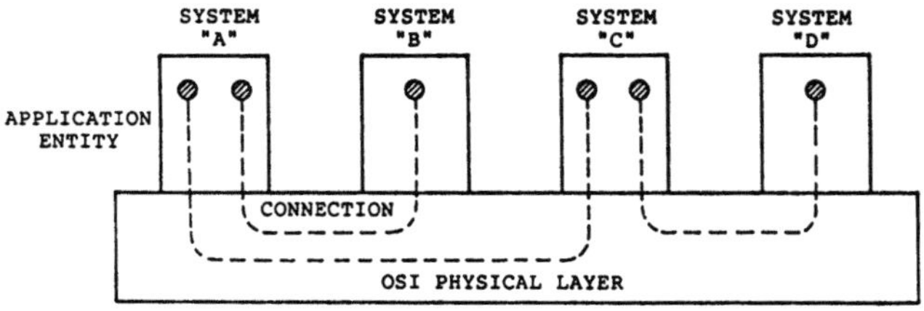

Fig. 1.8. Architecture of the Reference Model

The *system* means one or more computers with all the associated software, peripherals, terminals, application processes and also the human operators. The application process is a task-oriented activity executed for an application purpose.

The *application entity* is a portion of the application process dominant in the cooperation. The entity is a discrete, self-contained unit with all its properties defined. The entities are related to the layers. The entities of the peer layers communicate and cooperate with each other.

The entities create functional units that provide services for the next layer's entities and also requesting and receiving services for the lower layer's entities. Basically the entity is — in a wider sense — the general name of an object that is characterized by its properties; but the entity is also — in a narrow sense — the general name for things which are able to perforn information-exchange with their environment.

The *connection* establishes an association between the application entities. The *physical interconnection medium* means the model of the real medium that is suitable for the transmission of signals and messages.

Functional Division of the Distributed System

This functional division was a key decision in the OSI Reference Model (Fig.1.9).
The functions of the distributed system are divided into two parts:
— the data communication field,
— the data processing field.
Both fields have their own peculiar traditions that are preserved.

The data communication procedures use the tools and methods of the techniques of telecommunication. Data processing is strongly related to the operating system, to the code set of the terminals. The error detection of the data communication is generally based on the preservation (storing) of a copy and is based on a simple algorithm, but the error detection of the data processing is application-dependent.

The central concept of both parts is the communication. The data communication part deals with the traditional communication functions and the data processing part essentially reflects the nature of the communications between processes.

The functions inside the two basic layers of the Reference Model can be further divided and classified. This classification will be discussed later.

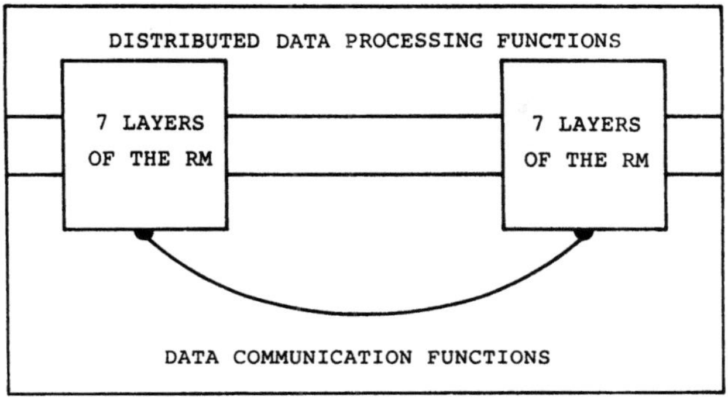

Fig. 1.9. Functions of the distributed system

Management in Open System Interconnection

The various management functions belong to three groups of open system interconnection:

— application management,
— system management,
— layer management.

The functions of application management are:

— the initialization of the parameters of the application processes,
— the initiation, maintenance and termination of the application processes,
— allocation and de-allocation of resources to application processes,
— detection and prevention of resource interference and deadlock,
— integrity and commitment control,
— security control,
— checkpointing and recovery control.

The system management functions are concerned with every layer, some characteristic functions are:

— activation and deactivation of resources,
— some program loading functions,
— initialization and modification of open system parameters,
— reporting status or status changes,
— reporting statistics,
— error detection and diagnostics,
— reconfiguration and restart.

The layer management is either an element of the system management or an integral part of a protocol.

The OSI management functions are shown in Fig. 1.10.

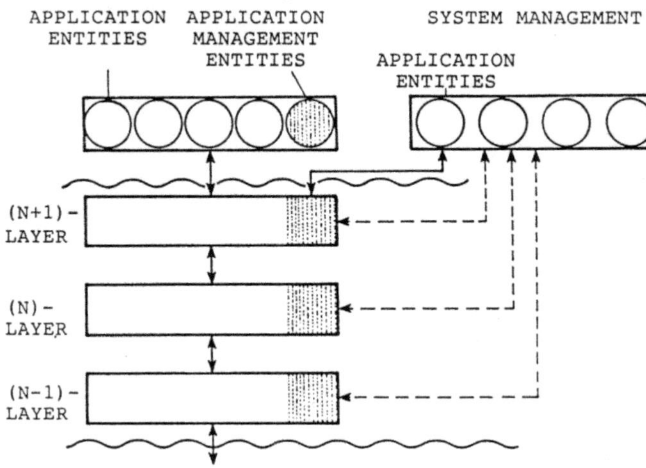

Fig. 1.10. OSI Management

1.3 LAYERS, LAYER TERMS AND LAYER ELEMENTS

The General Principles of Layering

Every system is composed of layers (Fig. 1.11). Specifying the interconnection of open system architectures means precisely defining the peer layer relationship of different systems and the relationship between the adjacent layers of a single system.

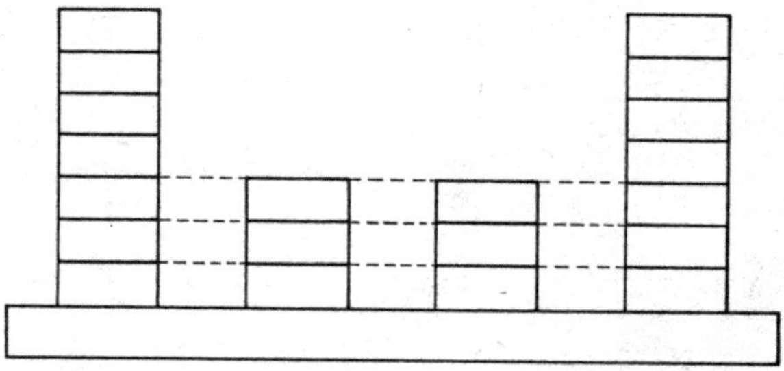

Fig. 1.11. The layered Reference Model

How should the layers be selected? Where should a layer boundary be placed, and how many boundaries should be created? In this respect, the following points are relevant:

— do not determine too many layers,
— the boundary should be well defined and the number of services and interactions across the boundary should be minimized,
— separate layers should be created to handle functions which are manifestly different, e.g. in the process performed or the technology involved,
— similar functions should be collected into the same layer,
— previously useful boundaries should be kept,
— separate layers for the different data abstractions should be created.

Layer Elements

Every layer of the Reference Model is composed of the same type of layer elements. The interpretation of the layer elements is shown in Fig. 1.12.

The layer elements are:
— entities,
— protocols,
— services,
— service access points.

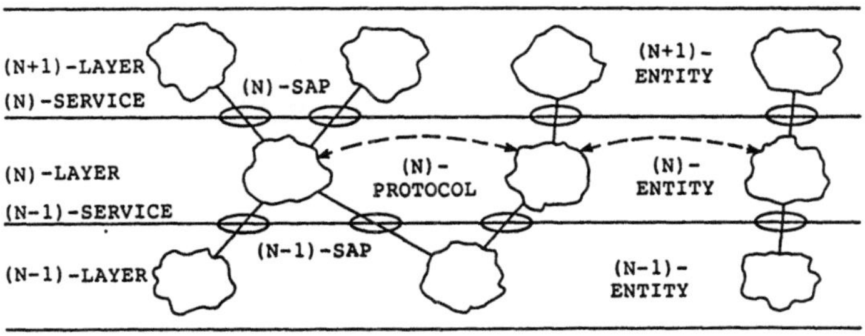

Fig. 1.12. The layer elements

We have learnt the concept of the application entity in Section 1.2. The layer entity is a particular functional unit of the layer. The (N)-entity completes the (N-1)-service received from the (N-1)-layer, and it adds to the (N)-service in the (N+1)-layer, through the service access point.

The (N)-protocol controls the cooperation among the (N)-entities. The protocol is the fundamental concept of the Reference Model, it defines the syntactic, semantic and temporal rules of cooperation. The protocol essentially defines the rules of network operation. The (N)-protocol describes the rules of cooperation among the entities, defining exactly the cooperation based on the (N-1)-services. The (N-1)-services fulfil the (N)-function. The (N)-functions increase the value of the (N-1)-service to provide an (N)-service to the (N+1)-entities.

The (N)-service is the capability of the (N)-layer which is provided for (N+1)-entities at the boundary between the (N)-layer and the (N+1)-layer.

The (N)-entities receive the (N)-services through the service access point (SAP). Briefly: the (N)-SAP is a logical interface between the (N)- and (N+1)-entity. Only one (N)-entity and one (N+1)-entity belong to one (N)-SAP, but one (N)-entity can establish connections through more (N)-SAP.

The entities and SAPs are referred to by identifiers. The (N)-title is the name of the (N)-entity. The title can be global or local. The (N)-address is the name of the (N)-SAP.

Identifiers. Naming and Addressing

The entities and interfaces can be referred to by identifiers. We can refer with names and titles to the entities and with an address to the assigned point of the interfaces (Fig. 1.13).

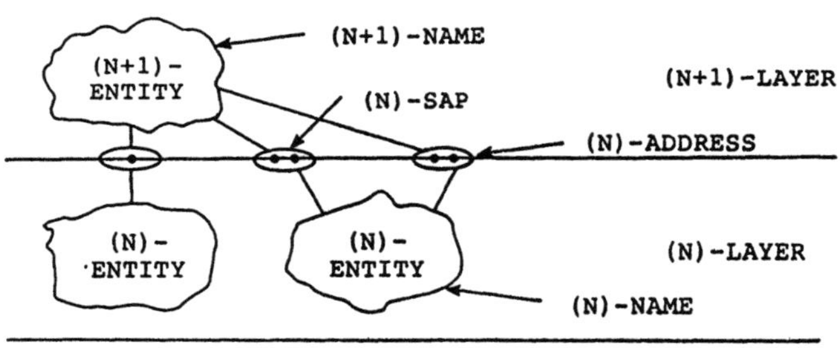

Fig. 1.13. Identifiers

The name is a linguistic object identifying an entity. The name can be global (it is valid in the whole network in other words universal), or local (it is valid only in a small region). The names have three types:

- the primitive name allocated to the entity by the denominator (it is forbidden to give the same primitive name inside a network),
- the descriptive name is a statement, which is valid only for one entity in the system,
- the general name is the common name of the entity set.

The name of the entity is the title.

Every service access point (SAP) has an address. The connection end point (CEP) is inside the service access point and has a special identifier named the connection endpoint identifier. The allocation of the entity names and SAP-addresses is possible with the directory. The mapping of the addresses are (Fig. 1.14)

— 1:1,
— hierarchical,
— tabular.

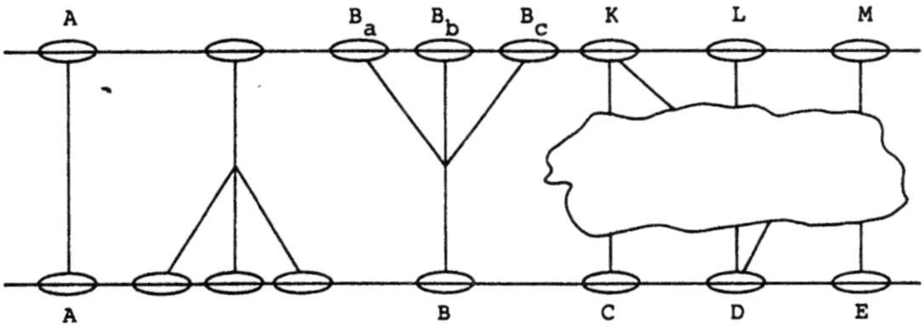

Fig. 1.14. Address mapping

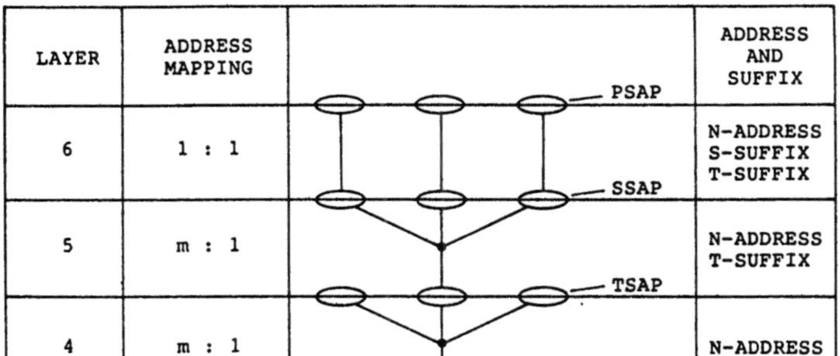

Fig. 1.15. Address and suffix in four layers

The hierarchical mapping differs from multiplexing. Multiplexing is effective for the connection establishment and hierarchical mapping is valid for the addressing. A suffix can also be given to the hierarchical mapping layer by layer besides the address. The (N)-suffix is part of the (N)-address and it is a single parameter inside the (N)-SAP completing the (N)-address. Figure 1.15 demonstrates the address mapping as well as the change of the address and suffix from the presentation layer to the network layer.

1.4 LAYER FUNCTIONS

The General Layer Functions

The (N)-function is a set of functions providing (N)-services to the (N+1)-layer executed by (N)-entities.

The layers are connected to the outside world through two interface surfaces

— the first interface is between the remote peer layers,

— the second interface is between adjacent layers.

The protocols determine the rules of coordination and interworking between the peer layers. These rules primarily deal with

— connection establishment and release,

— data transfer,

— error handling.

Data from the (N)-users carry the information among the (N)-entities. All (N)-user data has (N)-protocol control information to control the coordination among the (N)-entities.

The rules of the interface of the adjacent layers describe the connection with the upper or lower layer. Only the uppermost layer and the lowest one are exceptions, because of not having adjacent layers in both directions.

The information between the adjacent layers is called interface data, its basic measure is the interface data unit. The interface data unit is composed of the interface data and the interface control information. The (N)-interface control information coordinates the operation between the (N)-entity and the (N+1)-entity.

Connection Establishment and Release

If we want to establish a connection between remote (N)-entities, certain prerequisite conditions should be fulfilled:

— an active (N-1)-connection,

— both (N)-peer-entities are in a suitable state for connection establishment.

If an (N-1)-connection does not exist, it should first be established. Its prerequisite is an active (N-2)-connection. Therefore a connection can be established between any remote peer-entities, if every lower connection is established. If an (N)-entity is busy or it is not in a suitable state for establishing connection, it should be released, or it is necessary to wait until the entity is ready for establishing a new connection.

The connection can be a one-way connection (simplex), a two-way alternating one (half-duplex), or a two-way simultaneous one (duplex).

The connection can be released at once or with a delay. If a connection is interrupted directly after the release request, then the user data units between the two entities are lost. The connection is preserved by delayed release until the transport of the last data unit has finished.

The connections connect the CEPs of the adjacent layers. The ratio of the CEPs of the adjacent layers may be

The connections connect the CEPs of the adjacent layers. The ratio of the CEPs of the adjacent layers may be

— *1:1* this means that the numbers of CEPs of the adjacent layers are the same,
— *n:1* multiplexing; n is the number of the CEPs of the upper layer and the lower layer has only one CEP,
— *1:n* splitting; n is the number of CEPs of the lower layer and only one CEP belongs to them in the upper layer.

The three connection types are shown in Fig. 1.16.

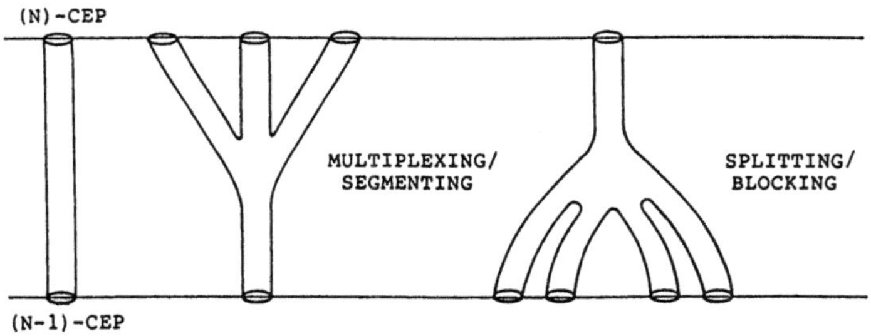

Fig. 1.16. Connections in the adjacent layers

The purposes of multiplexing are:
— more economical and more efficient use of the (N-1)-connection,
— several (N)-connections established where only a single (N)-connection exists.

The purposes of splitting are:
— higher reliability,
— higher transport performance, higher throughput.

Data Transfer. Data Units

The main task of the computer network is information exchange. There is a possibility to send short messages — generally one data unit — during connection establishment and release. Nevertheless, the information goes from the source in the network to the destination only during normal data transfer.

There are three types of data units:
— Protocol Data Unit (PDU),
— Interface Data Unit (IDU),
— Service Data Unit (SDU).

The structure of the first two data units can be seen in Fig. 1.17, the relationship between the second and third type of data unit is shown in Fig. 1.18.

The protocol data unit going between the (N)-layer entities is composed of two parts:

$$(N)\text{-PDU} = (N)\text{-PCI} + (N)\text{-UD}$$

where

(N)-PCI is the Protocol Control Information (PCI) of the (N)-layer which coordinates the information exchange between two (N)-entities using the (N-1)-connection(s), and

(N)-UD is the User Data (UD) passing between the (N)-entities.

	CONTROL	DATA	DATA UNIT
(N)-(N) ENTITY PAIRS	(N)-PCI	(N)-UD	(N)-PDU
(N)-(N-1) ADJACENT ENTITY PAIRS	(N-1)-ICI	(N-1)-ID	(N-1)-IDU

Fig. 1.17. Relationship between data units

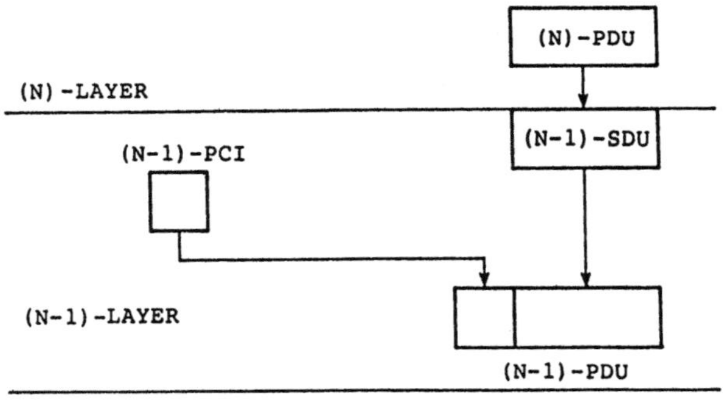

Fig. 1.18. Logical connection between data units of adjacent layers

The interface data unit crossing the boundary of the (N)- and (N+1)-layer is also composed of two parts:

$$(N)\text{-IDU} = (N)\text{-ICI} + (N)\text{-ID}$$

where

(N)-ICI is the Interface Control Information (ICI) coordinating the information exchange between the (N+l)-entity and (N)-entity,

(N)-ID means the transported Interface Data (ID) between the (N) and (N+1)-entities.

Figure 1.18 demonstrates how the (N)-PDU is mapped onto the (N)-SDU.

There is a possibility for expedited data transfer, but the length of the transported message is limited.

Data units are not necessarily the same size in the various layers. Therefore a single (N)-service data unit can be decomposed into many (N)-protocol data units, this being called *segmenting*, and many (N)-service data units can be formed into a single (N)-protocol data unit, this being called *blocking*. Both types of mapping are shown in Fig. 1.19.

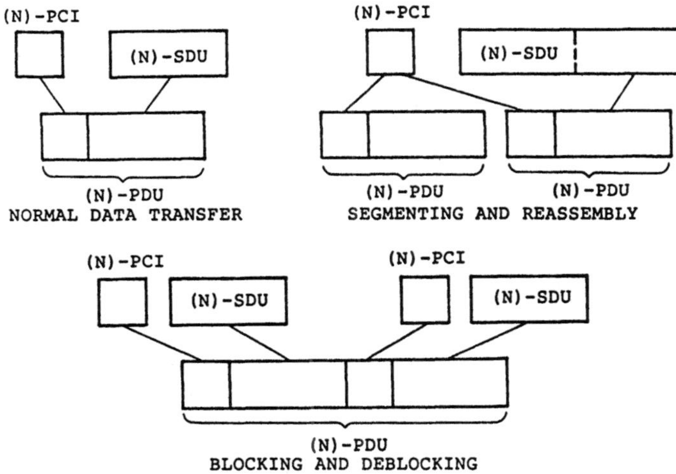

Fig. 1.19. Mapping of data units

A sequence number should be allocated to every data unit and this number used to replace the data units in the original order on reception. The sequence number belonging to a certain data unit changes from layer to layer.

The flow control is also a function of normal data transfer. The flow control restricts either the number of PDUs going between the peer-entities or the number of IDUs crossing adjacent layers.

Error Handling

The *acknowledgement* is the basis of error recognition, e.g. error detection by transmission. Every (N)-protocol data unit can be distinguished from all others as it has a unique identifier. In this way the receiver can inform the sender after the reception of each (N)-PDU. The acknowledgement should also detect if an expected data unit does not arrive. (N)-PCI belonging to the (N)-PDU contain the information necessary for the acknowledgement.

The most frequently occurring errors detected by the acknowledgement are the following:

— one data unit is lost,

— one data unit arrives twice,

— the segmenting or blocking of the data units happens in an erroneous sequence.

Error handling also includes detection of erroneous data. The reset function is the clearing of the connection of data units between corresponding (N)-entities and the restarting of flow control. The information part of the (N)-PDU contains an element starting the reset function between two communicating entities after loss of synchronization. The reset function resets the (N)-entities to a predefined state. The possibility of data loss and data duplication is increased by resetting.

1.5 LAYER CONVENTIONS

The Model of the Layer Services

For layer conventions, unified layer agreements are needed in every layer of the Reference Model so that unified concepts, terms and terminology can be used. The services provided by the layers are central to the layer conventions. The layer services are also defined by an abstract model. This abstract model represents the relationship of two opposite partners (Fig. 1.20):

— service user,

— service provider.

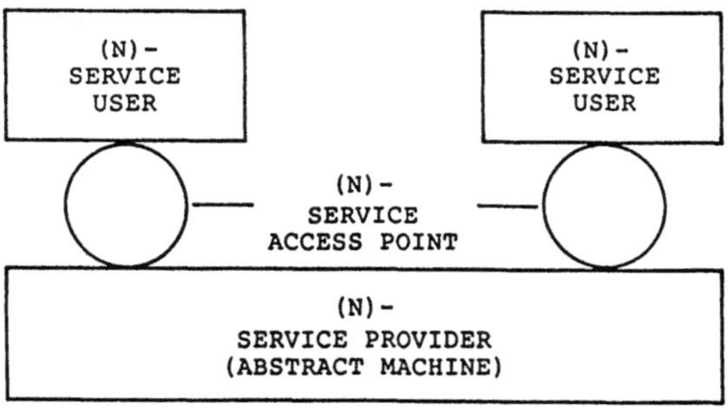

Fig. 1.20. Layer service model

The service user is the abstract representation of the number of all entities inside a single system that utilize the layer service through a single service access point. The service provider is an abstract machine modelling the number of all entities which provide the services together with the way in which the service users consider them. The interaction between the user and service provider is realized by service primitives.

Service Primitives

Service primitives are of two types:

— a unique primitive, which is a logically indivisible and temporary instance,

— a complex primitive, which can contain other service primitives; this is a logically divisible and transient event sequence.

The service primitives can be seen in Fig. 1.21. The service user can send primitives of two kinds:

— *request* which activates a procedure,

— *response* which announces the end of a procedure.

The service provider can send primitives of two kinds:

— *indication* which either activates a procedure or indicates that a procedure was activated by the service user at the service access point of the peer entity,

— *confirmation* which announces the end of a procedure.

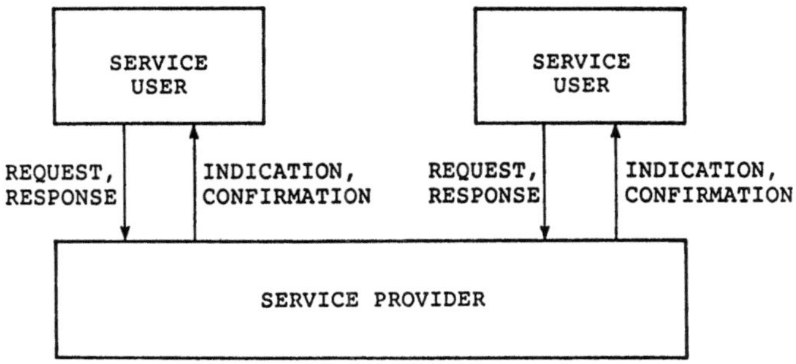

Fig. 1.21. Service primitives

Service primitives belonging to two service users should be considered carefully. The service primitives have parameters too. These parameters can be conditional. The notation of service primitives is composed of three parts according to the agreement:

— initial letters referring to the layer of the OSI model,

— collective name relating to the primitive group,

— individual name referring to the type of the primitive.

The standard initial letters are the following:

— A application layer,

— P presentation layer,

— S session layer,
— T transport layer,
— N network layer,
— D data link layer,
— PH physical layer.

The collective name relates to the nature of the procedure, e.g.

— establish,
— abort,
— connect.

The third part of the service primitive denotes the type of primitive, viz.

— request,
— response,
— indication,
— confirmation.

Time Sequence Diagrams

Time sequence diagrams demonstrate the relationship between the primitives and the order of the events (Fig. 1.22).

The two outside fields denote the two service users, the middle field represents the service provider; the vertical lines denote the time scale, giving the moments of time belonging to the service access points [GOTZ-85].

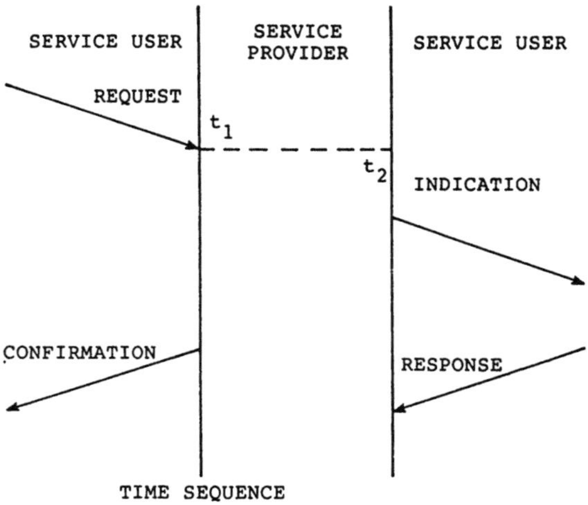

Fig. 1.22. Time sequence diagram

1.6 PROTOCOLS

The Basic Communication Concept of Open Systems: The Protocol

According to the ISO standard dealing with the interconnection of an Open System and the Reference Model:

> "the (N)-protocol is the set of syntactic and semantic rules determining the information exchange among the (N)-entities in such a way that the (N)-functions should be fulfilled".

The syntactic rules determine the format, e.g. the bit map of the messages; the semantic rules determine the procedures, and the correct and proper sequence of the dialogue. The semantic rules comprise a concealed way of specifying the timing and also the time specific characteristics.

The formats carry two types of information: viz. that for the data field, and, the information for control.

Data fields have always a control field, but control fields do not always have a data field. For example, the connection can be established or released only by means of messages comprising control information.

The main types of procedure are:

— connection establishment and release,
— data transfer,
— addressing,
— error handling,
— routing,
— flow control,
— assignment of resources.

Different names can be assigned to different procedures using the notation in Section 1.5 (Layer conventions).

The time specifications determine the time-out limits of waiting, i.e. the time interval within which the system accepts an answer. Therefore the protocol is defined by the formats, procedures and time behaviour.

The following data transfer errors can often be caused by protocol errors:

— a frame is lost,
— a frame is doubled,
— a frame arrives at a node with a bad address,
— the correct sequence of frames following each other is mixed up.

If the format is erroneous then either an undefined answer or an erroneous answer arises. If the procedure is erroneous, then an answer that is not permitted in the given state arrives. A timing error causes a lost frame or a message that is not replied to.

The general features of the protocols will be discussed again in Part III, but will then be based on the philosophy of finite state machines.

Protocol Features and Their Control

Protocol features can be classified according to several view points. One possibility is to classify them according to quantity and quality features. The quantity features can be given in bits, seconds, number of messages, percentages, etc. whereas the quality features characterize the reliability, the logical correctness as well as the identity of two protocols.

The most frequent quantity features are:

— length of data unit,
— window size,
— number of frames,
— segment size,
— number of data units per block,
— time duration of the connection establishment,
— time duration of the connection release,
— transfer time of the data unit,
— number of misdeliveries,
— number of detected errors,
— ratio between control information and data field.

There are many and complicated relationships among the quantity features in view of which they can be calculated only by means of simplification. Simulation often gives more help with designing than with calculation. The quantity features can be measured under operation by different protocol-analysers and -testers as well as by the built-in network management centre. The measurement of the quantity features forms a base for tuning the network. The protocol benchmarks, which are now performed in several institutes, give a good performance comparison.

The qualitative parameters of the protocols characterize the error-free operation (e.g. proper flow control in case of a node failure), the logical correctness (e.g. whether the protocol has a deadlock) as well as conformance and identity with the specification. Error-free operation is controlled by simulation in the course of designing and by measurements during implementation and operation. The logical correctness should be controlled only in designing, several verification methods are known for this.

One of the most important steps of implementation is the conformance test (testing the conformance with the specifications). Conformance testing examines all three components of the protocols: formats, procedures and timing. Protocol parameters and control methods are enumerated in Table 1.1.

Table 1.1. Control methods for protocol parameters

	Quantitative parameters		Qualitative parameters		
	Performance measures	Number of errors	Errorless	Logical correctness	Identity
Design	simulation	simulation	simulation	verification	-
Implementation	benchmark test	measurement	measurement	-	conformance testing
Operation	measurement	measurement	measurement	-	-

The Protocol and Its Environment

It is not possible to test the operation of a single protocol separately from its environment in an isolated way. The environment of the protocol is: the layer of the protocol itself and the adjacent layers. The (N)-protocol means the rules of communication between two entities in the (N)-layer [TAR-87d].

The parameters of the transfer path between two entities are:

— transmission speed,
— error characteristics,
— parameters of the queues,
— the information quantity to be transmitted,
— traffic in the opposite direction.

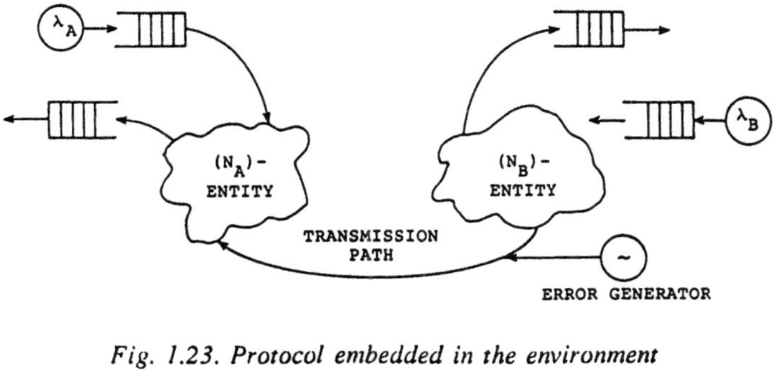

Fig. 1.23. Protocol embedded in the environment

The structure of the adjacent layers and their active condition also affect protocol operation. The adjacent layers can be described by active or passive models. In Fig. 1.23 the upper layer is characterized by an active model while the lower one is built into the transmission path. It is trivial that the error characteristics of the transmission path determine the number of acknowledgements of successful transmissions. Therefore, reasonably selected acknowledgement mechanisms are different in cases of different error percentages.

2. THE SEVEN LAYERS

2.1 ABOUT THE SEVEN LAYERS

The Seven Layers

In the standard recommendation of the Reference Model, ISO suggests the separation of the seven layers. The layers are as follows, starting with the lowest level:

— **Physical Layer**

This layer supports the transparent transfer of any bit sequence through the connection built into a physical communicating medium.

— **Data Link Layer**

It restricts the limits of the physical connection. It permits detectable errors and supports the re-establishment of the connection. It eliminates the quality errors of the transmission or, at least, restricts them.

— **Network Layer**

This layer ensures transparent data transfer, selects the appropriate path and routes the data through the selected path.

— **Transport Layer**

It establishes connection among the end-users, optimizes the usage of the resources depending on the character of the communication, and it allows the user to forget about matters of detail.

— **Session Layer**

It coordinates the cooperation between the communicating application processes.

— **Presentation Layer**

This layer transforms the data syntax into a form based on the requirements of the communicating application processes.

— **Application layer**

This layer specifies the nature of the communication required by the user and ensures that the tools helping the application processes reach the OSI environment.

The Functions of the Seven Layers

The main functions of the seven layers can be seen in Fig. 2.1.

(7) APPLICATION LAYER Application processes of the user, application-dependent services and management functions
(6) PRESENTATION LAYER Interpretation of data, transformation of formats and codes, encryption
(5) SESSION LAYER Administration and control of dialog between processes, segmenting, buffering, abnormal recovery
(4) TRANSPORT LAYER Transparent data transfer, end-to-end control, multiplexing, address mapping, packet or message assembly/disassembly, priority
(3) NETWORK LAYER Routing, flow control, connection establishment, segmenting, blocking, error recovery, network management
(2) DATA LINK LAYER Establishment, maintenance and release of data link connection, error recovery
(1) PHYSICAL LAYER Electrical, functional and mechanical interface

Fig. 2.1. The main functions of the seven layers

The lower layers (from the bottom upwards) are
— the physical layer,
— the data link layer,
— the network layer,
— the transport layer.
The upper layers are
— the session layer,
— the presentation layer,
— the application layer.

The main differences are that the communication functions in the lower layers are more concerned with data communication, while the upper layers are more concerned with data processing.

The problems of the upper layers are closely connected with the information representation. Different computer firms use different conventions for the information representation. They have different

- internal character codes,
- representation of numerical values,
- notation of the position of the bits,
- interpretation of the bits.

The different systems:
— printers,
— displays,
— text editors,
— file systems,
— data base systems,
— message handling systems,
— other application- and industry-oriented systems
cause the diversity of conventions in representing the information.

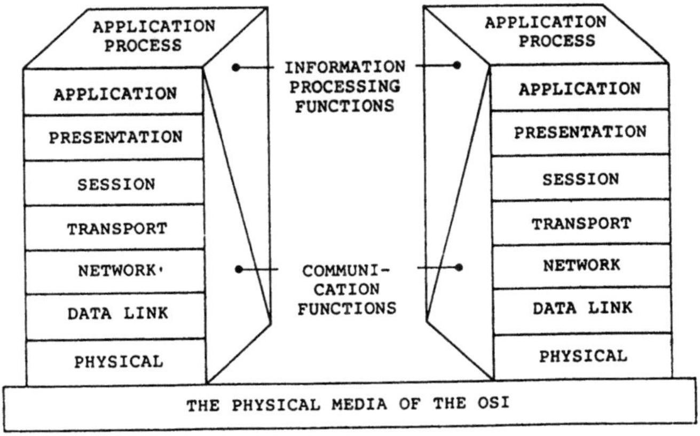

Fig. 2.2. Functions in the Reference Model

The distributed system is composed of the different systems and the distributed applications define the set of communication requirements. The lower layers deal with data transport between the end-systems. The data are represented as strings of binary octets whose quality is acceptable for the application processes.

The upper layers are responsible for the *interconnection* of application processes and for their dialogues. The interworking is the communication function to help the intelligent communication between the application processes and to execute the procedures necessary for information processing.

In brief, the Reference Model integrates the communication functions and the information processing functions. A very simplified illustration of the distributed functions was demonstrated in Fig. 1.9. A better approximation is in Fig. 2.2 whose is not so sharply defined [MEL-PET-84], [PU-PO-80].

2.2 THE PHYSICAL LAYER AND ITS PROTOCOLS

The Physical Layer and Its Services

The purpose of the physical layer is to provide the mechanical, electrical, functional and procedural characteristics used to activate, maintain and disconnect the physical link between the data-link layer entities transmitting the bits [BER-80].

What kind of services are provided between the physical layer and the data link layer?

— physical layer connections, in which bits pass between the data link entities,

— service data units of the physical layer; these are either used of "one bit"-units by serial transmission, or "more bit"-units by parallel transmission; the transmission can be duplex or half-duplex,

— physical connection among endpoints (between two endpoints by point-to-point connection, among many endpoints by multipoint connections),

— physical layer-connection identifier, which specifies unambiguously the connection for the data link entities,

— sequencing; the physical layer delivers the bits in the same order as they were sent,

— error indication; the errors detected in the physical layer are reported to the data link entities.

The physical layer has two functions:

— the activation and deactivation of the physical layer connection,

— the transmission of the physical service data unit.

A physical layer connection with endpoints can be seen in Fig. 2.3. The figure demonstrates how the abstract physical layer connection is separated from the physical connecting medium in the real environment. The media dependent characteristics of the interface of the physical layer and the physical media should be specified and these will be discussed later.

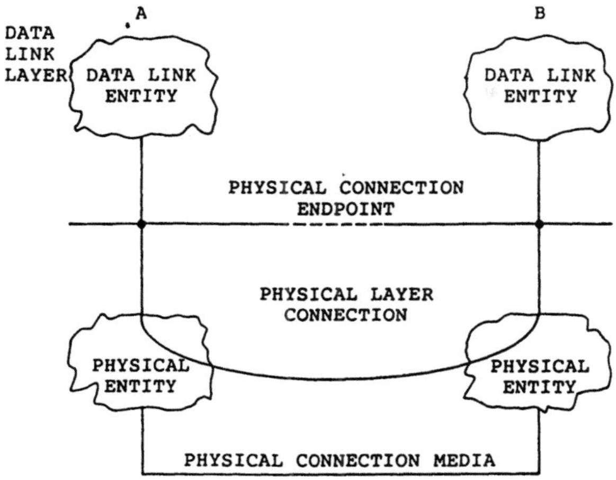

Fig. 2.3. Physical layer connection with two endpoints

The Data Transfer

The telecommunication net (sending or receiving messages by telephone or telegraph, either by radio signals or wires), or the public data network is the real physical connecting medium, e.g. here the actual data transfer takes place. If the data

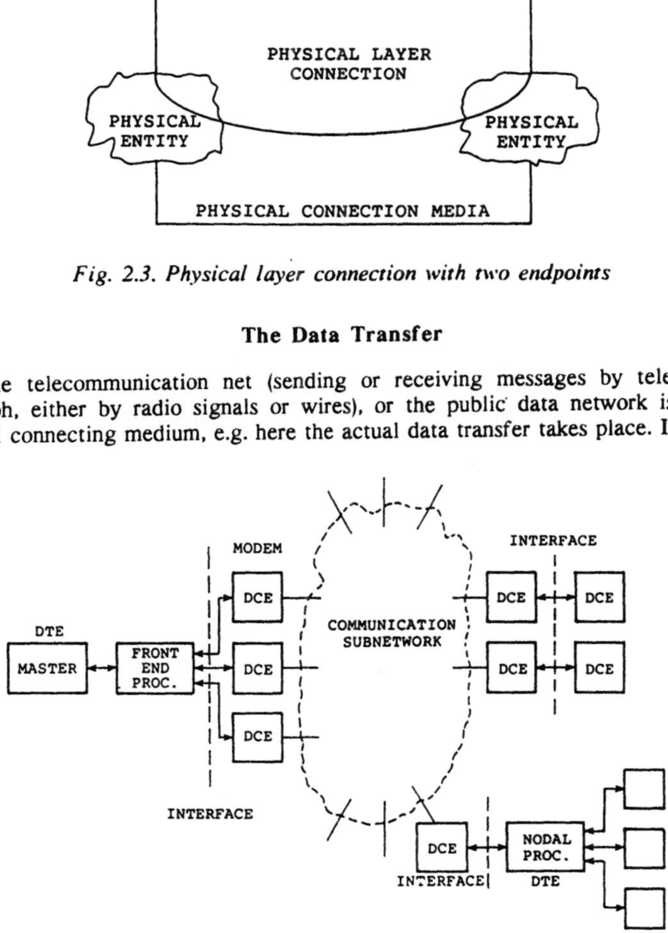

Fig. 2.4. The components of the data transmission

transfer is executed by telephone line, then the CCITT V.24 standard defines the rules of connection and if by a public data network it is defined by the X.21 standard. The data transmission shown in Fig. 2.4 is in a telephone net.

The basic units of data transmission are

— the Data-Circuit Terminating Equipment (DCE),

— the Data-Terminal Equipment (DTE).

The DCE is at the end of the circuit; the equipment is owned by the postal authorities and is connected to the DTE with a standard interface by the subscriber.

A typical DCE is a modem in a telephone network using the V.24 interface.

The DTE belongs to the user and it is connected to the data network. A DTE can be represented as a simple data station, from a terminal to a main frame.

The DCE is part of the communication subnetwork and is an integral part of the subnet. The interface between the DCE and DTE is the surface, whose mechanical, electrical, procedural and functional characteristics are specified in the standards.

Mechanical, Electrical, Procedural and Functional Characteristics of the Physical Layer

The DCE and DTE are connected by cables, the mechanical size and the number of pins are described by standards for the end of the cables and in the interfaces belonging either to the DCE or to the DTE. The connecting cable is either balanced or unbalanced. The balanced connecting cable should have a balanced generator sending positive and negative signals in relation to the ground signal. A negative signal represents binary 1, a positive signal represents binary 0. The amplitude is from 4-6 V. The receiver is a differential amplifier. The unbalanced interconnecting cable has an unbalanced generator producing negative signals with 4-6 V amplitude. Only the polarity of the signals is different by binary 0 and binary 1. The electrical characteristics of the balanced interconnection can be seen in Fig. 2.5.

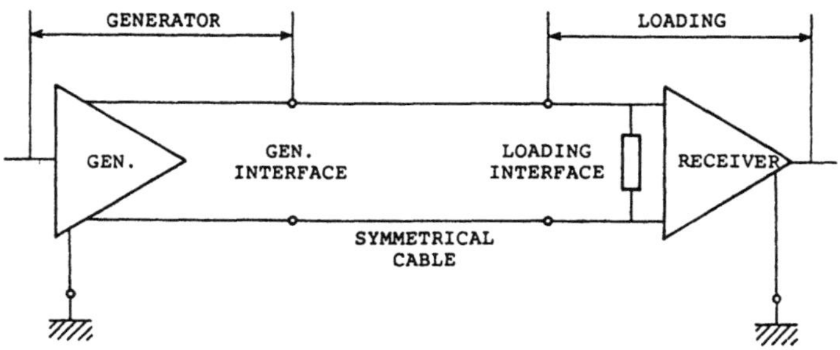

Fig. 2.5. Electrical elements of X.27 recommendation

The functional characteristics deal with the data, control, timing and grounds. The functional elements of the DTE/DCE interface of the X.21 recommendation are shown in Fig. 2.6.

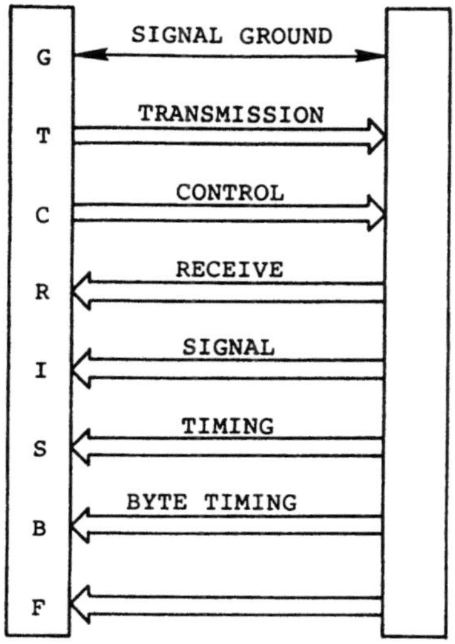

Fig. 2.6. Functional elements of X.21 recommendation

The procedural characteristics of the X.21 recommendation are, for example, the quiescent signal of DCE and DTE:

— DCE ready,
— DCE not ready,
— DTE ready,
— DTE not ready.

The procedural characteristics specify the establishment of maintenance test loops and the error detection.

2.3 THE DATA LINK LAYER AND ITS PROTOCOLS

The Data Link Layer and Its Services

The objective of the data link layer is the detection and possible correction of any errors occurring in the physical layer.

What kind of services does the data link layer provide for the network layer?

— data link connection establishment between two network entities,
— the error-free transport of the data link service data units,
— error notification,
— data link disconnection,
— recovery from certain error states during the data link connection.

The difference between the high reliability of computers and the substantially lower quality of the transmission path causes problems in the data link layer. Reliable and effective transmission between two nodes should be ensured in spite of unreliable transmission media, noisy channels and finite transmission time.

The main tools of this are

— the packing of the data into frames,
— the application of error detecting codes,
— the acknowledgement of received information,
— the sequence numbering,
— the window mechanism.

Any kind of binary sequences can realize the protocols in the layers above the data link layer, but the recognition of bit errors in a continuous, optionally long sequence is impossible. Therefore the bit-sequence is divided into discrete pieces called *frames*. The beginning and the end of the frame are marked to help recognition. Sometimes the length of the frame is given too. The packed data in frames is forwarded and it is padded out with redundant data for error diagnostics. Either the erroneous sequence is recognizable (error detecting coding) or the original sequence can be restored (error correcting coding) based on the redundancy algorithm. If we have applied error detecting codes, then the erroneous frame must be repeated. The principle of the error recognition is generally based on the division of the data of one block with a predefined bit-sequence and the remainder is delivered with the data block. The division is repeated by the reception and then the equality of the calculated and transmitted remainders is compared.

The *acknowledgement* of the received data increases the reliability of the transmission. A positive acknowledgement indicates the correct reception of the message or the frame. The copy of the message is stored on the source station until the arrival of a positive acknowledgement. A negative acknowledgement indicates an erroneous reception or a lost frame. If we wait till the arrival of a positive acknowledgement (individual acknowledgement) after sending a frame, then the traffic will be extremely slow therefore it is necessary to introduce a more efficient acknowledgement instead of the individual acknowledgement. One means of efficient acknowledgement is to acknowledge the received frame in the header of the information (or control) frame coming from the opposite directions. Frames should have sequence numbers in order to refer to them in the acknowledgement. *Sequencing* can be broadened for any range of natural integers, but from a practical point of view it is not very useful to apply too big a number. The sequencing helps to limit the number of unacknowledged packets. The *window* is equal to the number of unacknowledged frames or packets.

Generally the frames are numbered from 0 to n-1, if n is the highest sequence number. The sequencing is modulo n. The zero frame comes after the (n-1)th frame.

Data Link Functions

These are the functions in the data link layer:
— data link connection establishment and release; this connection is built on existing, already activated physical connections,
— the mapping of the data link service units to data link protocol units,
— the splitting of one data link connection into more physical connections,
— the framing of the bits; this is also called the synchronization of the data units,
— the control of the sequencing,
— the detection of transmission errors, erroneous format or operation errors; the error can originate from the physical layer connection or from the erroneous operation of the peer entity,
— the recovery after a detected error or informing the network entities about errors from which a recovery is impossible,
— flow control,
— the identification of the data link entities and parameter exchange.

Classification of the Data Link Protocols

One of the main data link functions is that of packing the bits into frames. The classification of the data link layer protocols is based on the frame structure. Essentially every frame, independent of its class, has the following structure:

control part	information	control part	error code

What is the difference between the frame structures in Fig. 2.7?
— character-oriented protocol; the control part is composed of control characters,
— bit-oriented protocol; the beginning and the end of the frame are denoted by flags,
— character-number-oriented protocol; the length of the data field should be given in advance.

The character-oriented protocols are issued at the same time as the data communication. Nowadays, character-oriented protocols are frequently used. Binary Synchronous Communications (BSC) will be discussed in detail in Section 2.4.

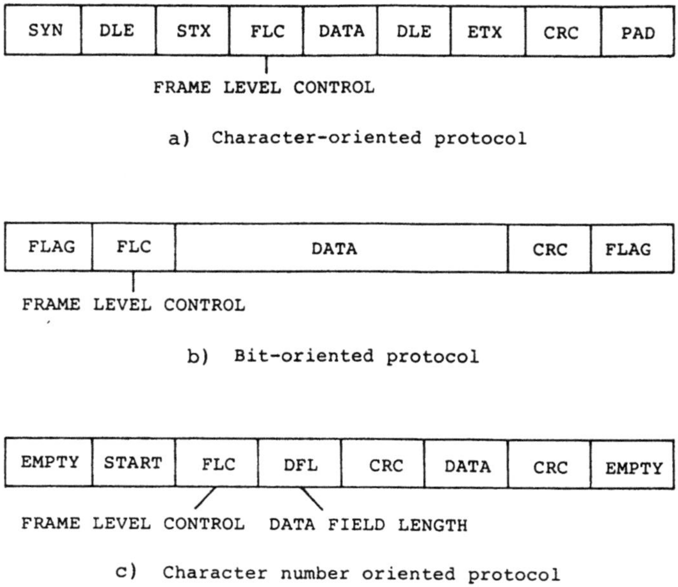

Fig. 2.7. Frame types of data link protocols

Why was it necessary to develop protocols with other structures?

— the hardware and software implementation is difficult because there are data and control characters inside a code set,

— transparency can be ensured only by means of complicated methods due to the character-oriented structure,

— the full-duplex possibilities are not efficiently utilized,

— it is suitable primarily for batch operations.

The general use of interactive applications and the request for more efficient connections initiated the development of bit-oriented protocols. These are better adapted to the different applications and configurations because they are code-independent and therefore transparent. Their basic mode is the duplex connection and for this reason the traffic is more economical and efficient. The High-Level Data Link Control (HDLC) protocol will be discussed as an example in Section 2.5.

2.4 BINARY SYNCHRONOUS COMMUNICATION (BSC) PROTOCOL

The Main Feature of BSC

The Binary Synchronous Communication (BSC) is a byte-oriented, logically half-duplex data link control protocol. Its main feature is that the establishment and maintenance of the synchronization between two stations is ensured by the sending and recognition of a well-defined, synchronizing character-sequence.

Data link communication can be realized as a point-to-point connection or a multi-point one. By multi-point connection one station (the primary or master) takes the dominant part on the line, this station controls the complete communication, inquiring about and selecting any other station. The data communication is always between two stations, the others in the network are in a passive observing state.

The characters in the transmission code set of the protocol may be

— graphic (numeric, alphabetic, special),
— functional (Horizontal Tabulator, DELete),
— data link control (Start of Text, End Of Transmission)

characters.

The protocol has three standard code sets:

— EBCDIC,
— USASCII,
— Six-Bit Transcode.

A station can ask for the right to the data link control at a point-to-point connection by sending a SYN(ENQ) sequence. Furthermore we denote the initial synchronizing sequence composed of 3-9 synchronous characters as SYN.

With a centralized multi-point connection the right for the data link control always belongs to the master station which invites every other station one after another, with a defined sequence, to send (polling) and to receive (selecting).

The delivery of the messages is in blocks, in order to realize effective error-diagnostics. The block is initiated with a Start of Text (STX) character, it is followed by the text, by the information part, and the block is terminated either with an End of Intermediate Transmission Block (ITB), with an End of Text Block (ETB) or with an End of Text (ETX) character. If the transmission of any other control information is necessary, then a header can be sent before the STX character. The header begins with a Start of Header (SOH) character. A special character can follow the SOH character in order to ensure data-protection and to allow interpretation of the header.

Various error diagnostic procedures ensure the detection of transmission errors in the BSC protocol. If the USASCII line code is used, then the error detection of the characters is performed by the generation and control of odd parity bits and the error detection of the blocks is ensured by Block Check Characters (BCC). The Block Check Characters can be either Longitudinal Redundancy Checking Codes or Cyclic Redundancy Checking (CRC) Codes. The sender sends the BCC characters after the block terminating character. The receiver executes the determined error diagnostic procedure, character by character, and compares the generated error checking codes with the arrived error checking codes produced by the sender. If the two codes are different, the receiver replies to the sender with a Negative Acknowledge (NAK), to indicate that the block arrived erroneously, and should be sent again.

The receiver acknowledges every received and completed block (terminated with ETB or ETX). The acknowledgement occurs with ACKO/ACKl sequences after an error-free block, and with a NAK after an erroneous one. If the acknowledgement does not arrived in time, or if it is incorrect, the sender then asks the receiver to repeat its answer with a SYN (ENQUIRY) sequence. After have finished sending the data a SYN (EOT) sequence is sent to the receiver by the sender.

In the Sixbit Transcode the CRC-12 block checking code is used, and in the EBCDIC line code the CRC-16 block check is applied.

The Character Set of BSC

The protocol has the following data link control characters and sequences:

DLE/EOT	Data Link Escape (End Of Transmission)
SYN	Synchronous character to establish and maintain synchronization
SOH	Start Of Header, the initiator character of the block header
STX	Start of Text, the initiator character of the block information part and the terminator character of the header, if any
ITB	End of Intermediate Transmission Block. The sender also sends a BCC character after the ITB, but there is no change in the direction of the transmission therefore the receiver sends the acknowledge either after an ETB or after an ETX
ETB	End of Text Block
ETX	End of Text, block terminator characters. The sender forwards one or more BCC characters after these characters, then it changes the direction and waits for the receiver's acknowledgement
EOT	End Of Transmission character denotes the end of the transmission
ACKO/ACK1	Positive acknowledgement sequence; the receiver has received the block without any errors
WACK	Positive acknowledgement sequence; sent by the receiver if it is temporarily unable to receive any more data. The sender answers with ENQ on the WACK sequence and the receiver repeats the WACK sequence from time to time until it becomes ready to receive. The repetition time depends on the time-out value of continuation
NAK	Negative Acknowledge The receiver sends NAK after observing an error on block reception
DLE	Data Link Escape Character

RVI Reverse Interrupt positive acknowledge
The receiver asks the sender for a temporary exchange of functions in order to send an urgent message. This sequence can only be repeated after an ENQ response by the sender

TTD Temporary Text Delay
The sequence is composed of STX/ENQ and sent by the sender to the receiver. This sequence detects that the sender can-not continue the sending. The receiver answers with NAK in the TTD sequence. The sender repeats the TTD sequence until it is able to finish sending. The repetition time depends on the time-out value of continuation.

```
EXAMPLES

a/ Normal data transfer

   SENDER                                          RECEIVER

SYN/ENQ

   line reversing
                                                   SYN /ACKO/

   line reversing

SYN/STX/ /block 1/ /ITB/ /BCC

SYN/block 2/ /ITB/ /BCC/

.
.
.
SYN/block n/ /ETB//BCC/

   line reversing
                                                  SYN /ACK 1/

   line reversing

SYN /last block/ /ETX/ /BCC/

   line reversing
                                                  SYN /ACKO/

   line reversing

SYN /EOT/

b/ WACK dialogue

   SENDER                                         RECEIVER

   .
   .
   .
```

```
SYN/STX /block n/ /ETB/ /BCC
                                         SYN /WACK/
SYN /ENQ/                                SYN /WACK/
SYN /ENQ/                                SYN /ACK1/
SYN /STX//last block/ /ETX//BCC/
   :                                     SYNC /ACKO/
c/ RV1 dialogue
   SENDER                                RECEIVER
   :
SYN/STX//block n//ETB//BCC/
                                         SYN /RV1/
SYN/STX//next block//ETB//BCC/
SYN/EOT/                                 SYN /ACKO/
                                         SYN /ENQ/
SYN /ACKO/
   :                                     SYN/STX//block1//ETB//BCC/
d/ TTD dialogue
   SENDER                                RECEIVER
   :
SYN/STX//block n//ETB//BCC/
SYN/STX//ENQ/                            SYN  /ACKO/
                                         SYN  /NAK/
SYN/STX//next block//ETB//BCC/
   :
```

Time-out

In the BSC protocol the time-out values are interpreted as:

a) *Sender* time-out: 1 sec
The sender inserts into the text every second, either two synchronous characters following each other by normal transfer or a DLE/SYN sequence by transparent transfer in order to maintain the synchronization.

b) *Receiver* time-out: 3 sec
The sender waits 3 seconds for the answer of the receiver. If the response does not arrive during this time, or the response is not appropriate, then the sender tries to compel the receiver, with an ENQ, to give an answer.

c) *Connection* release time-out: 20 sec

If a station detects an interruption longer than 20 seconds during the traffic by a circuit-switched network, then the station disconnects.

d) *Time-out* of continuation: 2 sec

The receiver/sender repeats every 2 seconds the given sequence under the WACK and TTD dialogue until the ability to receive/send is re-established.

The BSC protocol makes transparent transmission possible. In this case a DLE character is sent before all data link control characters, except the ENQ character, the SOH character and the response sequences. It should be noted that the BSC generation by transparent transmission is different from its generation by normal transfer.

2.5 HIGH-LEVEL DATA LINK CONTROL (HDLC) PROTOCOL

The Fundamental Concept of HDLC

ISO worked out the HDLC protocol and suggested it as a standard. At present it is the most widely used bit-oriented protocol. What kind of requirements should be fulfilled?

— the beginning and the end of the information should be marked,
— a difference should be made between the sender and the receiver,
— errors should be detected,
— the sequence should be kept,
— many stations should be able to use the same line,
— it should be code-independent (to ensure transparency),
— it should be adaptable in different environments (batch, interactive, half-duplex, duplex, point-to-point, multipoint),
— it should be efficient (the ratio of information transfer to control transfer should be large),
— it should have a fast recovery after error detection,
— there should be high reliability.

The HDLC operation modes are unbalanced or balanced. The *unbalanced* configuration contains a master and a slave station, the most frequently used methods are the point-to-point connection and the dropped one (Fig. 2.8).

The responses are normal or asynchronous in unbalanced mode. The slave station can send messages in *Normal Response Mode* (NRM), when the master station calls upon it. The call is executed with a poll bit. If the master sets the value of the poll to one, the slave can send messages. Once the slave has finished the message sending then it gives back control to the master with the final bit. The poll bit and the final bit are in the same position in every HDLC frame, the interpretation depends on the sender. If the sender is the master, it is the poll bit. If the sender is the slave, it is the final bit.

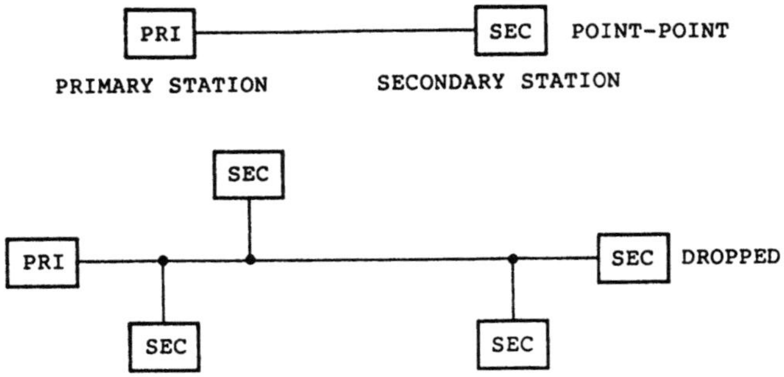

Fig. 2.8. Asymmetrical connections

In *Asynchronous Response Mode* (ARM) the control depends on the appointed master station, but the slave can also initiate a connection. The slave is not obliged to give back control after finishing the transmission.

In a *balanced* configuration the sender and the receiver are of the same rank, i.e. either both sides are master and slave or both sides are combined stations (Fig. 2.9). The response is always asynchronous. This mode is known as Asynchronous Balanced Mode (ABM). None of the stations waits for permission from the other. Rather the poll and final bits acknowledge messages.

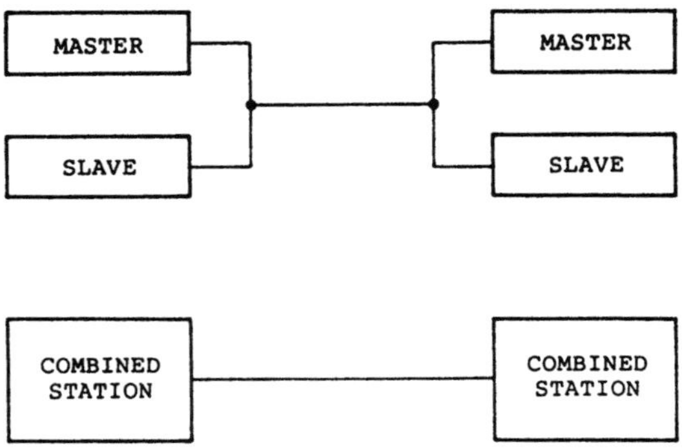

Fig. 2.9. Asynchronous Balanced Mode

In the **X.25 CCITT** recommendation an asynchronous balanced protocol is in the data link layer, it is denoted as **LAPB** (Link Access Procedure Balanced).

Frame Types

In an unbalanced configuration the master sends commands and the slave sends responses. In a balanced configuration the combined stations both send and receive commands and responses.

The frames are of three types:

— Information frame,
— Supervisory frame,
— Unnumbered frame.

The information frame has the following structure:

FLAG	ADDRESS	CONTROL	INFORMATION	FCS	FLAG
8 bits	8 bits	8 bits		16 bits	8 bits

All frame types — information frames as well — in HDLC start and end with a flag. The flag is the following:

0 1 1 1 1 1 1 0 .

The address in a command is the address of the destination, in the response it is the address of the source. In the standard recommendation the suggested addresses are

1 1 0 0 0 0 0 0 (command)
1 0 0 0 0 0 0 0 (response).

If the message to be sent is a circular, addressed to every node, then a global address composed of ones should be written.

The next 8-bit part is the control field. The first bit of an information frame is always zero.

The structure of the frame is

1	2 3 4	5	6 7 8
0	N (S)	P/F	N(R)

where

N(S)	is the sequence number of the sender
N(R)	is the sequence number of the receiver
P/F	poll bit/final bit.

The sender sequence number is the sequence number (modulo 8) of the most recently sent information frame, the receiver sequence number is the sequence number of the following information frame which is awaited, but has not arrived. Therefore the receiver sequence number is a hidden acknowledgement.

The length of the control field is 16 bits, when 127 messages can start without acknowledgement. In this case 7 bits are necessary for defining the sender and receiver sequence number. Thus the control structure in modulo 128 is

0	1 2 3 4 5 6 7 8	9	10 11 12 13 14 15 16
0	N (S)	P/F	N(R)

This is the extended mode.

The information field follows the control field. The length of the information parameter is a system parameter and its value is set by agreement between the users.

The Frame Check Sequence (FCS) terminates the information field. FCS is a 16-bit-sequence in LAPB.

The generator polynomial equation is

$$x^{16} + x^{12} + x^{5} + 1$$

The *supervisory frames* make up the second group of HDLC frames. The supervisory frames control data transfer and ensure recovery after data loss in I-frame errors.

The general form of the supervisory frame is:

FLAG	ADDRESS	CONTROL	FCS	FLAG
8 bits	8 bits	8 bits	16 bits	8 bits

The structure of the supervisory frame is

	1	2	3	4	5	6 7 8
RR	1	0	0	0	P/F	N(R)
RNR	1	0	1	0	P/F	N(R)
REJ	1	0	0	1	P/F	N(R)
SREJ	1	0	1	1	P/F	N(R)

The first bit of every supervisory frame is 1, the second bit is 0, the fifth- bit is the poll/final and the last 3 bits describe the N(R) which is a hidden piggybacking acknowledgement. The third and fourth bits denote the differences among the supervisory functions, these are the following:

— RR is Receive Ready

- detects the readiness for reception
- releases the busy situation after an RNR transmission
- it acknowledges the earlier received frames until the sequence number N(R-1)

— RNR Receive Not Ready

• it detects a busy state

— REJ Reject

• the information frames should be repeated from the N(R) given in the reject frame

— SREJ Selective Reject

• only one information frame should be repeated which has the N(R) sequence number in the SREJ-frame.

The unnumbered frames belong to the third group of HDLC frames. They are responsible for the connection establishment and release, and for unnumbered acknowledgement. They have the same general format as the supervisory frames.

The unnumbered frames have the following control field structure:

	1	2	3	4	5	6	7	8
SABM	1	1	1	1	P/F	1	0	0
DISC	1	1	0	0	P/F	0	1	0
UA	1	1	0	0	P/F	1	1	0
FRMR	1	1	1	0	P/F	0	0	1

The types of unnumbered frames are the following:

— SABM Set Asynchronous Balanced Mode; it sets the addressed station in asynchronous balanced mode

— DISC Disconnection; it stops the previous mode

— UA Unnumbered Acknowledge; it acknowledges the reception and acceptance

— FRMR Frame Reject; it detects errors which cannot be eliminated with a repetition of the same frame. For example invalid command, I-frame longer than permitted, invalid N(R).

The Disconnected Mode (DM) also belongs to the mandatory command/response set of the LAPB. Its interpretation is that the DM-sender is logically released from the data connection and is in the disconnected phase. The DM-response informs the other station that mode commands cannot be started and cannot be settled.

Table 2.1 gives a summary of the control fields of all three frame types:

In the table, S represents the supervisory bits and M is a modifier bit.

The above stated are valid for the (modulo-8) not-extended mode. In extended mode (modulo 128) the N(S) and N(R) have seven bits and the base mode can be settled with the SABME command. E means "extended" at the end of the word. The bitmap of the U-frame is the same in extended mode.

Table 2.1. The control fields of the LAPB frames

	1	2	3	4	5	6	7	8
Information I-frame	0		N(S)		P/F		N(R)	
Supervisory S-frame	1	0	S	S	P/F		N(R)	
Unnumbered U-frame	1	1	M	M	P/F	M	M	

Error-free Operation of the HDLC

A normal and errorless operation has three phases:
— connection establishment,
— information transfer and
— release.
The connection establishment and release are demonstrated below

```
A → B SABM;
B → A         UA;
```

a) Asynchronous balanced

```
A → B SNRM,P;         DISC,P;
B → A         UA,F;           UA,F;
```

b) Normal response mode, unbalanced
Half-duplex transmission
Information transfer
Only B wants to send information

```
A → B SNRM,P;        RR(0),P;                                  RR(3),P;
B → A         UA,F;           I(0,0); I(1,0); I(2,0)F;                  I(3,0); I(4,0);
```

c) Normal response mode, unbalanced
Half-duplex transmission
Information transfer
Only A wants to send information

```
A → B SNRM,P;      I(0,0); I(1,0),P;         I(2,0); I(3,0);
B → A          UA,F;                RR(2),F;
```

d) Normal response mode, unbalanced
Half-duplex transmission
Information transfer
Both station, A and B send messages

```
A → B I(0,0); I(1,0); I(2,0),P;                    I(3,2),P;        I(4,3);
B → A                            I(0,3); I(1,3),F;           I(2,4),F;
```

The first two dialogues contain the connection establishment phase, too. The sender and receiver sequence number is given in brackets after the commands and responses. The letter P denotes the poll bit, the letter F denotes the final bit.

The information exchange is easier to understand by half-duplex transmission, but the full-duplex transmission makes the dialogue more efficient.

The following examples illustrate the NRM duplex transmission:

a) Only A sends I-frames

```
A → B SNRM,P;       I(0,0),P; I(1,0);  I(2,0),P; I(3,0); I(4,0),P; I(5,0);
B → A          UA,F;         RR(1),F;            RR(3),F;          RR(5),F;
```

b) Only A sends I-frames
The number of poll bits is decreased

```
A → B SNRM,P;       I(0,0),P; I(1,0); I(2,0); I(3,0); I(4,0); I(5,0),P;
B → A          UA,F;         RR(1),F;                                  RR(6),F;
```

c) Normal response mode, unbalanced
Duplex transmission
A and B send I-frames

```
A → B SNRM,P;      I(0,0),P; I(1,0); I(2,1); I(3,2); I(4,3); I(5,4); I(6,5);
B → A          UA,F;    I(0,0); I(1,1); I(2,2); I(3,3); I(4,4); I(5,5);
```

The information exchange is illustrated with two types of acknowledgement

a) Asynchronous response mode, balanced
Duplex transmission
Information transfer
Piggy-backed acknowledgement

```
A → B I(0,0),P; I(1,0); RR(1),F; I(2,2),P; I(3,3); I(4,4),F;
B → A     I(0,0),P; I(1,1),F;    I(2,2); I(3,2); I(4,3),F;
```

b) Asynchronous response mode, balanced

Duplex transmission

Information transfer

Acknowledgement with **RR**

A → B I(0,0),P; I(1,0),P; RR(1),F; I(2,1),P;

B → A I(0,0),P; RR(1),F; I(1,2); RR(2); I(2,2),P;

The alteration of the acknowledgement time depends on the business of the other station in the normal case:

a) **B** does not want to send information, it acknowledges with **RR**

A → B I(i,1);

B → A RR(i+1);

$\longleftrightarrow$

T_{ACK} ,

b) **B** sends just then an I-frame when the message from **A** arrives and **B** does not want to send a new I-frame

A → B I(i,1);

B → A I(1,i); RR(i+1);

$\longleftrightarrow$

T_{ACK}

c) **B** sends just then an I-frame when the message from **A** arrives and **B** wants to send the next I-frame

A → B I(i,1);

B → A I(1,i); I(2,i+1);

$\longleftrightarrow$

T_{ACK}

The above demonstrated examples surveyed the phases of connection establishment information exchange and release of the normal error-free operation.

Just one more question should be examined. What affects the performance of the HDLC procedure in an error-free case?

— the modulus of the sequence number,

— the usage of **P/F** bits.

The modulus can be 8 or 128 in HDLC. A larger modulus ensures the transmission of more short messages than a smaller one. The performance is the opposite for longer messages.

The usage of **P/F** bits may lead to unnecessary expectation when waiting for a poll bit from the master station. The suggested strategy is to send a **P**=1 bit as often as possible. It is much more favourable in balanced mode because both stations can initiate the next step.

Erroneous Operation of the HDLC

The station sending an I-frame starts a clock and the clock stops after the arrival of the acknowledgement. The clock starts only at a frame with P=1. The acknowledgement should arrive within the time-out. If not, the sender repeats the sending of the I-frame.

The other tool for error detection is the negative acknowledgement. The receiver sends an REJ response, if the arrived I-frame has an erroneous sequence number. The REJ response contains the sequence number of the last correctly accepted frame. It repeats which frame is awaited. The sender repeats the sending of every frame with equal sequence number or bigger than the sequence number of the incorrect I-frame.

a) In the next example the I-frame of the A station with sequence number 1 is out of order.
  Asynchronous balanced mode
  Duplex transmission
  The double underlining means a repeated message.

```
A → B I(0,0); I(1,0); I(2,2); I(1,2);
B → A I(0,0); I(1,1);         a repeated frame
```

b) The SNRM starting the connection establishment is incorrect .
  Normal balanced mode
  Duplex transmission

```
A → B SNRM,P;  time-out  SNRM,P;        I(0,0);
B → A                            UA,F;
```

c) The SABM starting the connection establishment and UA acknowledgement is incorrect
  Asynchronous balanced mode
  Half-duplex transmission

```
A:
A → B SABM;  time-out  SABM;      time-out SABM;
B → A                       UA;                   UA;
```

d) The I-frame is incorrect, the negative acknowledgement is with the N(R) of the opposite I-frame

Asynchronous balanced mode
Half-duplex transmission

```
A → B I(0,0); I(1,0); I(2,0),P;                  I(1,1); I(2,1); I(3,1),P;
B → A                           I(0,1),F;                                  I(1,4),F;
```

e) The I-frame is incorrect, the negative acknowledgement is with the N(R) of the opposite RR-frame

Asynchronous balanced mode
Full-duplex transmission

```
A → B I(0,0),P; I(1,0);   RR(1),F; I(2,1),P; RR(2);                 I(1,3),P;
B → A    I(0,0),P   RR(1),F; I(1,1),P;         I(2,1),P; RR(1);            RR(2),F;
```

f) The next two dialogues demonstrate the difference between REJ and SREJ by incorrect I-frame

Normal response mode
Full-duplex transmission
The effect of REJ:

```
A → B I(0,0),P; I(1,0); I(2,0); I(3,1); I(4,2);      I(2,3); I(3,4); I(4,5);
B → A                I(0,1); I(1,2); I(2,2); REJ(2); I(3,2); I(4,2);
```

The effect of SREJ:

```
A → B I(0,0),P; I(1,0); I(2,0); I(3,1); I(4,2);           I(2,3); I(5,4);
B → A              I(0,1); I(1,2); I(2,2);   SREJ(2); I(3,2); I(4,2); I(5,5);
```

g) The next dialogue demonstrates the effect of REJ by incorrect I-frame, but the mode is balanced

Asynchronous balanced mode
Full duplex connection

```
A → B I(0,0); I(1,0); I(2,2); I(3,3);   I(1,3); I(2,4); I(3,5); I(4,6);
B → A I(0,0); I(1,1); I(2,1);  REJ(1); I(3,1); I(4,1); I(5,2); I(6,3);
```

h) The repeated I-frame becomes incorrect therefore no answer arrives within the time-out. The sender calls upon the receiver with RR(6) for acknowledgement. The sender repeats I(i) and I(i+1) after receiving RR(6).

```
A → B I(i,1); I(i+1,1); I(i+2,2); I(i+3,3); I(i,3);              RR(6),P;
B → A   I(1,i);      I(2,i); REJ(i);      I(3,i); I(4,i); I(5,i);          RR(i),F;
```

2.6 THE NETWORK LAYER AND ITS PROTOCOLS

The Network Layer Services

The network layer establishes, maintains and disconnects network connections between the communicating application entities and supports the service data unit exchange between the transport entities. It makes transparent the flow control, the switching modes and the data link connection to the transport entities.

What kinds of services are provided for the transport layer by the network layer?

— conveying the transport data transparently,
— allocating network addresses to the transport entities,
— establishing the network connection,
— supporting the network connection end point identifier,
— delivering the network service data units (expedited delivery also exists),
— controlling the flow,
— numbering the sequence of packets,
— resetting.

Routing is the primary function inside the network layer. It will be discussed on page 56.

Typical network functions are the multiplexing of network connections and the segmenting and blocking of data units. The network layer disassembles and assembles the transport messages into packets. It sends control information about its state to the other network layers. It recognizes the message priorities and sends the messages according to their priority.

X.25 Packet Level Procedure

The CCITT X.25 recommendation deals with the information exchange and protocol rules of the three lower layers.

The X.25 packet level procedure summarizes the rules valid for the third layer. At present only one packet may be put in the information field of an I-frame. In the future it is very likely that more packets will be permitted in one frame because of the increasing number of traffic requests.

The virtual call connects two DTEs during the call. For every virtual call a logical channel group sequence number is allocated and also a logical channel sequence number during the connection establishment phase, in order to maintain many virtual calls simultaneously. There is a maximum of 15 groups and 255 logical channels. The range of logical channel sequence numbers and the logical channel group sequence numbers allocatable to the virtual calls should be arrived at by a bilateral agreement, valid for a determined length of time.

The CCITT permits the establishment of a constant virtual circuit. It means that two DTEs are constantly connected through the appropriate logical channels. The network is responsible for the correct sequencing of the packets.

The datagram is an option among the packet-type services, when the user can send independent messages anywhere and can receive them from anywhere.

What are the basic elements of the X.25 packet level procedure?

— connection establishment and release,
— data transfer,
— interrupt (the interrupt-packet is responsible for the resolution of the blocked data transfer),
— flow control,
— resetting (it clears the logical channel of data and restarts the flow control),
— restarting (it disconnects all virtual calls at the same time and resets the constant virtual circuits allocated to the given DTE).

The data packets should be sequenced modulo 8 or modulo 128 according to the window size. The packets also deliver the sender sequence number P(S).

The object of a bilateral agreement in a system is that the packet sequence numbering is modulo 8 or modulo 128. The maximum length of a data field is 128 octets. A "more data" detecting bit is in the packets to provide for coherent transfer of data in amounts greater than 128 octets. The value of the "more data" bit is ONE if the packet to be transmitted has a direct continuation packet after it. The value of the "more data" bit is ZERO for the last packet and for an independent packet.

Packet Formats

The packets carry data, i.e. information, or control.

Packet types:

- Connection establishment and release
  - — Call Request
  - — Call Connected
  - — Disconnect Request
  - — Disconnect Acknowledgement
- Data and interrupt
  - — Data
  - — Interrupt
  - — Interrupt Acknowledgement
- Flow control and reset
  - — Receive Ready
  - — Receive Not Ready
  - — Reset Request
  - — Reset Acknowledgement
- Restarting
  - — Restart Request
  - — Restart Acknowledgement.

The general packet format is:

<table>
<tr><th></th><th>8</th><th>7</th><th>6</th><th>5</th><th>4</th><th>3</th><th>2</th><th>1</th></tr>
<tr><td rowspan="2">common packet header</td><td colspan="4">General format identifier</td><td colspan="4">Logical group number</td></tr>
<tr><td colspan="8">Logical channel number</td></tr>
<tr><td></td><td colspan="7">control</td><td>"more data" bit</td></tr>
<tr><td></td><td colspan="8">Further packet header and/or user data</td></tr>
</table>

The header of the data packet:

<table>
<tr><th>8</th><th>7</th><th>6</th><th>5</th><th>4</th><th>3</th><th>2</th><th>1</th></tr>
<tr><td>Q</td><td>D</td><td>0</td><td>1</td><td colspan="4">Logical group number</td></tr>
<tr><td colspan="8">Logical channel number</td></tr>
<tr><td colspan="3">P(R)</td><td>M</td><td colspan="3">P(S)</td><td>0</td></tr>
<tr><td colspan="8">Data</td></tr>
</table>

Q is the qualifier bit in the general format identifier. The whole packet series can be put in two levels depending on the value of Q, D is the delivery acknowledgement bit, M is the "more data" bit.

The packet format of the call request:

octets	bits 8 7 6 5	bits 4 3 2 1
1	General format identifier	Logical group number
2	Logical channel number	
3	Packet type identifier	
4	Address length of calling DTE	Address length of called DTE
the number of octets increases dynamically	DTE addresses	
	0 0	The length of the facilities
	Facilities	
	User data	

Routing

An important objective of the network layer is the routing. What kind of requirements are needed in the routing?

— to ensure quick and error-free message transfer,

— to be adaptable to the network topology changing, because of node and channel errors,

— to be adaptable to the altering source-destination traffic.

In routing there are deterministic and stochastic methods. The most frequently used deterministic routings are the following:

— flooding,

— fixed,

— NRCC (Network Routing Control Centre).

Stochastic routing methods are:

— random,

— isolated.

The greatest safety is provided by flooding-routing. Every message goes to every node. A node sends the message further after determining that the given node has not delivered the same message earlier and that this node is not the destination.

Fixed routing allocates a well-defined route to every node pair. All nodes have a routing table determining the next node to each possible destination. Figure 2.10 provides a good illustration of fixed routing. (See Table 2.2.)

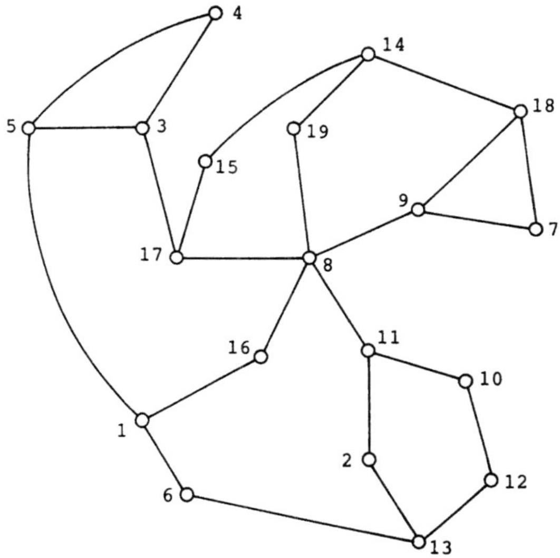

Fig. 2.10. Fixed routing

The Network Routing Control Centre is a special node. This node collects the traffic data and at regular intervals calculates a routing table for every node.

Random routing has a more practical realization. Every method has the common principle that the selection of the appropriate route is based on some kind of random distribution of the business of nodes and channels belonging to the possible routes. The next node or the whole path can be suggested.

The routing table in the eighth node can be seen in Table 2.2.

The isolated or delay-table routing stores the expected value of the delay for every node depending on the channel selected. The channels with the minimum delay are described in vector form. The delay table and the channel vector of the minimum delay are demonstrated in Table 2.3.

Table 2.2. Routing table

Destination	Next node
1	16
2	11
3	17
4	17
5	17
6	16
7	9
x	x
9	9
10	11
11	11
12	11
13	11
14	19
15	17
16	16
17	17
18	9
19	19

Table 2.3. Delay table and channel vector

destination \ channel	L1	L2	L3	L4
N1	7	1	8	5
N2	7	4	3	5
.	.	.	.	.
.	.	.	.	.
.	.	.	.	.
NI	4	9	8	5
.	.	.	.	.
.	.	.	.	.
.	.	.	.	.
NN	2	3	6	5

channel vector
L2
L3
.
.
.
L1
.
.
.
L1

2.7 THE TRANSPORT LAYER AND ITS PROTOCOLS

The Transport Layer and the Transport Services

The transport layer is an interface between the bottom three data-communication layers and the top three data-processing layers. The objective of the transport layer is to support the transparent data transport among the transport entities and to optimize the usage of the available communication resources.

What kind of services are provided for the session layer by the transport layer?

— connection establishment,
— data transfer (normal and expedited),
— release.

The functions of the transport layer are

— transmission of TPDUs;
— multiplexing and demultiplexing; to share a single network connection between two or more transport connections;
— error detection;
— to detect the loss, corruption, duplication, misordering or misdelivery of TPDUs;

— error recovery
  to recover from detected and signalled errors;
— connection establishment
  to match the TS-users' requested quality of service with the services offered by the network layer;
— concatenation and separation to collect several TDPUs into a single NSDU at the sending transport entity and to separate the TPDUs for the receiving transport entity;
— segmenting and reassembling
  to segment a single data TSDU into multiple TPDUs at the sending transport entity and to reassemble them into their original format for the receiving transport entity;
— splitting and recombining
  to allow the simultaneous use of two or more network connections to support the same transport connection;
— flow control
  to regulate the flow of TPDUs between two transport entities on one transport connection;
— transport connection identification to identify uniquely a transport connection between the pair of transport entities supporting the connection during the lifetime of the transport connection;
— expedited data transfer to bypass the flow control of normal data TPDU
— TSDU delimiting
  to determine the beginning and ending of a TSDU.

The transport service primitives are summarized in Table 2.4, their directions and positions between the layers can be seen in Fig. 2.11.

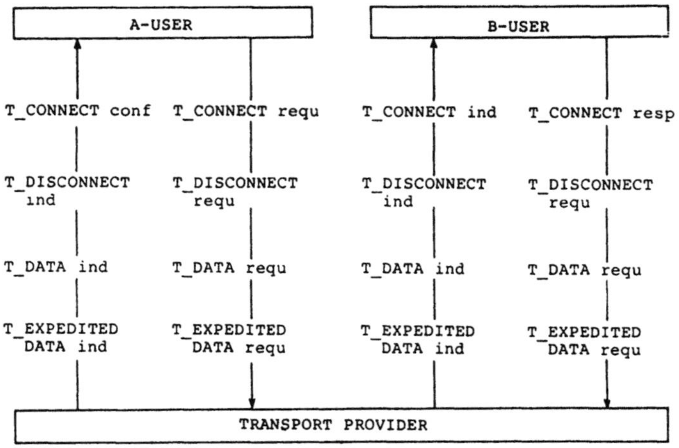

Fig. 2.11. Transport primitives

Table 2.4. Transport service primitives

PRIMITIVE		PARAMETERS
T-CONNECT	request indication	Called Address Calling Address Expedited Data option Quality of services TS-User Data
T-CONNECT	response confirm	Responding Address Quality of services Expedited Data Option TS-User Data
T-DATA	request indication	TS-User Data
T-EXPEDITED DATA	request indication	TS-User Data
T-DISCONNECT	request	TS-User-Data
T-DISCONNECT	indication	Disconnect reason TS-User-Data

The Table 2.5 describes the transport protocol data units and their parameters.

Every TPDU has a length indicator at the beginning of its structure and the CR, CC, DR, DT and ED have a data field for user data.

Table 2.5. Transport protocol data units

TPDU	PARAMETERS	
	Fixed part	Variable part
Connection Request (CR)	CR code	TSAP-ID
	Credit	TPDU-size
	Destination-reference	Version number
	Source-reference	Security parameter
	Class and option	Checksum
		Additional option selection
		Alternative protocol class
		Acknowledge time
		Throughput
		Residual error rate
		Priority
		Transit delay
		Reassignment time
Connection Confirm (CC)	CC-code	as CR
	Credit	
	Destination-reference	
	Source-reference	
	Class and option	

Table 2.5. (continued)

TPDU	PARAMETERS	
	Fixed part	Variable part
Disconnect Request (DR)	DR-code Destination-reference Source-reference Reason	Additional information for clearing Checksum
Disconnect Confirm (DC)	DC-code Destination-reference Source-reference	Checksum
Data (DT)	DT-code Destination-reference TPDU-NR EOT	Checksum
Expedited Data (ED)	ED-code Destination-reference ED-TPDU-NR EOT	

Table 2.5. (continued)

TPDU	PARAMETERS	
	Fixed part	Variable part
Data acknowledgement (AK)	AK-code Destination-reference Your-Turn-Number	Checksum Subsequence number Flow control confirmation
Expedited Data Acknowledgement (EA)	EA-code Destination-reference Your-ED-Turn-Number	Checksum
Reject (RJ)	RJ-code Credit Destination-reference Your-Turn-Number	
TPDU Error (ER)	ER-code Destination reference Reject cause	Invalid TPDU Checksum

Transport protocol classes

The transport services are unified, but five classes of the transport protocols have been developed to ensure the unified transport services independently of the network quality.

The functions of the transport layer have been organized into classes and options. A class defines a set of functions. Options define those functions within a class which may or may not be used.

The transport protocol classes:

a) Class 0; The Simple Class.
b) Class 1; The Basic Error Recovery Class.
c) Class 2; The Multiplexing Class.
d) Class 3; The Error Recovery and Multiplexing Class.
e) Class 4; The Error Detection and Recovery Class.

How can the network services be classified in terms of quality?

a) Type A: Network connections with acceptable residual error* rates and acceptable rates of signalled errors.
b) Type B: Network connections with acceptable residual error rates but unacceptable rates of signalled errors.
c) Type C: Network connections with unacceptable residual error rates.

The next question is which class of transport protocol should be used in conjunction with a given network connection.

Class 0: The Simple Class

It provides the simplest type of transport connection and is fully compatible with the CCITT recommendation S.70 for teletext terminals.

Class 0 has been designed to be used with type A network connections.

Class 1: The Basic Error Recovery Class

It provides a basic transport connection with minimal overheads. The main purpose of the class is to recover from network disconnects or resets. The Class 1 transport protocols provide transport services with correct sequence and correct data integrity in the case of detected errors (N-DISCONNECT, N-RESET).

This class has been designed to be used with type B network connections.

Class 2: The Multiplexing Class

It provides a way to multiplex several transport connections into a single network connection. This class has been designed to be used with type A network connections.

There are two options:

- use of explicit flow control,
- non-use of explicit flow control.

Class 3: The Error Recovery and Multiplexing Class

Class 3 has the characteristics of Class 2 plus the ability to recover from network disconnects or resets.

Class 3 has been designed to be used with type B network connections.

* Residual errors are all kinds of violation of the integrity of delivered data that are not observed or detected by the network layer

Class 4: The Error Detection and Recovery Class

It has the characteristics of Class 3, plus the detection of errors which occur as a result of the low grade of services available from the NS-provider.

The kind of errors that are detectable include:

— TPDU loss,
— TPDU delivery out of sequence,
— TPDU duplication and
— TPDU corruption.

These errors may affect control TPDUs as well as data TPDUs.

Class 4 has been designed to be used with type C network connections.

General remarks: Class 0 and Class 1 are known as CCITT-Classes, Classes 2, 3 and 4 are ISO-Classes because of their origin. It is worthwhile mentioning that the US National Bureau of Standards standardizes only Classes 2 and 4 [DAN-BRE-75], [DAN-BRE-76], [KOV-87].

2.8 THE SESSION LAYER AND ITS PROTOCOLS

The Session Layer

The purpose of the session layer is to provide the means necessary for cooperating presentation entities to organize and to synchronize their dialogue and to manage their data exchange. To do this, the session layer provides services to establish a session connection between two presentation entities and supports their orderly data exchange interactions.

A presentation entity has only two ways "to speak" with another entity:

— either it initiates a session connection, or
— it accepts a session connection.

The session layer is missing from many computer network architectures. There are many systems having the same session and transport address, but the address mapping between the session and transport layers can be either one to one or n to one. A simplified explanation of the session layer is that it ensures the connection of application processes and their data exchange. If the lower layers are working well then some presentation and application functions help to fulfil this interworking.

What kind of session services are provided for the presentation layer?

— session-connection establishment,
— session-connection release,
— normal data exchange,
— quarantine service (the destination can access a data unit only after special permission from the source),
— expedited data exchange,
— interaction management (this deals with the dialogue control),

— session-connection synchronization,
— exception reporting (exceptional states are observed and this information is passed to higher levels).

The session-service definition is described in ISO document IS/8326 and the session-protocol definition in IS/8326.

Session Layer Functions

The session layer is the lowest layer recognizing the presence of the users in the network. This layer provides the establishment, organization and release of the session communication. The processes running in different systems are coordinated in the session layer.

The tools of coordination are
— synchronization and
— checkpointing.

The session-layer functions are carried out by session entities in order to provide the session services for the presentation layer.

The session layer functions are:
- — establishment and release of the session establishment
- — mapping of session connections onto transport connections
  - one transport connection onto several consecutive session connections
  - several consecutive transport connections onto one session connection
- — data exchange through the session connection
  - normal data exchange
  - expedited data exchange
  - management data exchange
- — session connection and method control
  - selection of the dialogue type
  - dialogue control with a token
- — session-connection recovery
  - synchronization
  - resynchronization
  - reset
- — session-layer management.

Operation of the Session Layer

The first step of the operation is the *connection establishment*. The connection establishment is an agreement between two application processes concerning how the controlled dialogue should be realized in order to exchange data.

Any application process can initiate the connection establishment either at the user's request or as an answer to any other application processes request.

Naturally, an active connection of the lower layers is assumed. Many lower layer connections may belong to a single session connection. It is somewhat like a chess-game: a game of chess lasts from the first move until the last move, but a phone connection is active at one time only.

At the time the connection is established, the type of dialogue should be selected. The release can be normal, initiated by the application process or it can be abnormal, released because of the occurrence of an error.

The second step is the *data exchange*. This means the delivery of data and protocol control information between the end-systems. There are normal, expedited and special transparent data exchanges. Data exchange is executed according to the selected dialogue type. The types of dialogues are:

— one-way communication
  the data can advance only in one direction,
  e.g. the measured data is collected and furthered to a measurement centre
— two-way simultaneous communication
  the data can go in both directions at the same time,
  e.g. a time-critical control process
— two-way alternating
  the communication is active either in one direction or in the opposite direction,
  e.g. a data base inquiry system.

The *dialogue control* occurs with the aid of tokens. The token is a mechanism for determining the rights of sending through the session connection at a given instant.

During the information exchange the data flow is synchronized. The reason for this *synchronization* is for data flow and reset, so as to allow the return to a synchronization point after error recovery.

The synchronization mechanism has major and minor markers. The major markers are the main synchronization points. If a major marker is confirmed by the receiver, it becomes the reference point and the previous major markers are no longer valid. The minor markers are additional points between two major markers. Every marker has a sequence number. The same sequence number never occurs twice during a session connection.

2.9 THE PRESENTATION LAYER AND ITS PROTOCOLS

The Presentation Layer and Its Services

The purpose of the presentation layer is to take care of syntactically correct information exchange between communicating entities.

If heterogeneous systems are required for interworking, then the data elements in each system are locally represented according to several definitions. The definitions referring to the information representation are called the syntax and each system has its own local syntax. The interworking systems may have different local syntaxes, only the definite syntax rules describing the coding of the information exchange are the subject of a mandatory agreement.

This definite syntax is the transfer syntax.

The purpose of the presentation standards are to establish an agreement in the transfer syntax. This agreement is limited, e.g. 5-bit-code information can-not be represented as EBCDIC-codes.

The presentation layer has the following services:

— data transformation (code- and character-set transformation),
— data formatting,
— syntax selection,
— special transformation (encryption, compression, etc.).

The functions of the presentation layer are the following:

— session establishment request including transfer syntax negotiation,
— regeneration of transfer syntax,
— transfer of data,
— session termination request.

There is a one-to-one correspondence between presentation address and session address. There is no multiplexing or splitting in the presentation layer.

Transfer Syntax and Presentation Context

The information to be delivered has three types of syntax:

— source syntax,
— destination syntax,
— transfer syntax.

The source and destination syntax are local ones, the tranfer syntax characterizes the data crossing the session connection.

The transfer syntax selection can be executed either by name or based on a preliminary agreement. If the selection is by name, then the two entities should find a mutually acceptable transfer syntax.

If the transfer syntax selection is based on a preliminary agreement, then the two systems should recognize each other after the connection establishment.

The speciality of the presentation layer is the handling of presentation context. There is always an initial context enabling the exchange of protocol control information and transparent information in the presentation layer. Further contexts are defined by the P-CONTEXT-DEFINE service element.

The P-Context describes the mechanism for negotiating the transfer syntax — as is shown by the following example:

```
              System A                        System B

P-CONNECT:    X/EBCDIC ⎫
              X/ASCII  ⎬ ——→
              X/IA5    ⎭
                                 ⎧ P-CONNECT-confirm
                          ←——    ⎨   X/EBCDIC
                                 ⎩   X/ASCII
P-SELECT:     X/ASCII     ——→         .
                          ←——      P-SELECT-confirm
```

The active context is set to X/ASCII on both sides.

The transfer syntax often deals with recursively defined structures having structure elements, e.g. records whose elements are also records, trees or files. It is therefore important to define an appropriate syntactical notation. The CCITT has developed X.409 for a Message Handling System; ISO recommends Abstract Syntax Notation One (ASN.1) based on X.409. This permits representation of data elements with the following data types:

— primitives (without internal structure),
— constructions (compound, e.g. composed of one or more different components).

2.10 THE APPLICATION LAYER AND ITS PROTOCOLS

The Application Layer Services and Functions

The purpose of the application layer is to open a window in the OSI environment for the user. Every important event and all significant information are observable through this window. The user communicates with the open system through the application layer by means of which the user can inform the open system about the required user dependent parameters.

What kind of application services are provided for the user?

— to identify the communicating partners by name, by address, by definite description and by generic description,
— to list the entities that are available for communication,
— to check authority,
— to calculate the cost,
— to determine the service quality (response time, tolerable error rates),
— to identify any constraints on data syntax (character sets, data structure),
— to transfer information.

Which functions do the application layer contain?
— Logging on,
— Password checking,
— File transfer,
— Downline loading,
— Database access,
— Graphic procedures,
— Colour control,
— Table and figure editing,
— Virtual terminal,
— Job manipulation,
— Remote job access,
— Message handling sytems,
— Job-specific applications (word processing, electronic financial transaction, reservations on a plane, transaction processing).

The Application Process

The application process is one of the most important, in terms of the upper layers. The applications exist in the real world of the user, outside of the computer system. The application processes model the applications within the computer (Fig. 2.12).

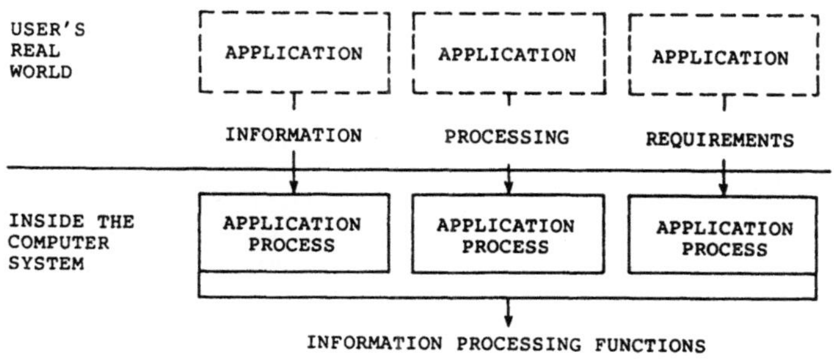

Fig. 2.12. Applications and application processes

Going into a little more detail about the application process, we can state that the application informally describes the information processing and that the application process formulates it formally. This formal description (Fig. 2.13) is composed of two parts:
— abstract representation,
— concrete representation.

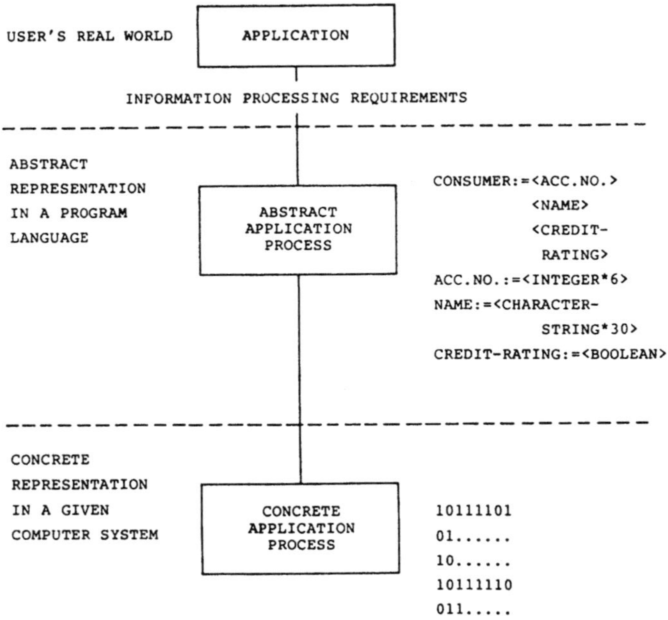

Fig. 2.13. The detailed interpretation of an application process

The *abstract representation* formulates the information processing functions in a program language. The data types are abstract categories, the procedures are defined according to the abstract syntax rules of the programming language. Why is it an abstract representation? Because the definition of data elements and procedures is independent of the realization in a given computer system.

The *concrete representation* is in a binary format-coded abstract representation.

The coding takes place in accordance with the concrete syntax of the computer system. The procedures used in this representation are the sequences of executable instructions. The binary coded character strings are described in ASCII, EBCDIC or other codes, the integers are binary numbers or two's complement, the negative values can be expressed with a sign-bit or with one's or two's complement. In this manner, the application process is compiled into concrete formats and is linked with system routines to create an executable module.

The interworking of the application processes means that two or more application processes execute a distributed information processing task. The task can be simple (e.g. the transmission of a character string) or complex. With a complex task it is necessary to synchronize many application processes.

Many application layer standards are needed to allow the interworking of many different kinds of application process. One part of these standards has a general validity (e.g. file transfer, access and management, virtual terminal service, job transfer and manipulation), the other part is valid only for special applications, e.g. a Message Handling System, interbanking, trade). A list of application layer standards can be found at the end of this chapter.

General Application Layer Protocols

There are protocols for general applications that were defined before the Reference Model was created. The following protocols are the most important:

— File Transfer, Access and Management (FTAM). This is responsible for file transfers between distributed computer systems and for the availability of the file content.

— Virtual Terminal Service (VTS). This protocol ensures computer access through a terminal.

— Job Transfer and Manipulation (JTM). This is responsible for the job execution in the distributed system.

Figure 2.14 surveys the application protocols.

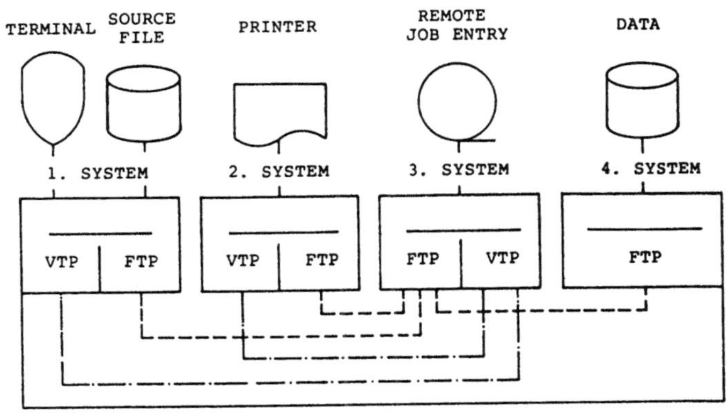

Fig. 2.14. Application protocols

The *File Transfer, Access* and *Management* Protocol is based on virtual file stores. A virtual file store is an abstract object with many attributes, which can be found in the majority of the real file systems. The attribute is a non-separable property of an object or a phenomenon, which cannot exist or cannot imagine without this property.

The basic types of file transfer (Fig. 2.15) are

— the file store,

— file output,

— to run a remote file.

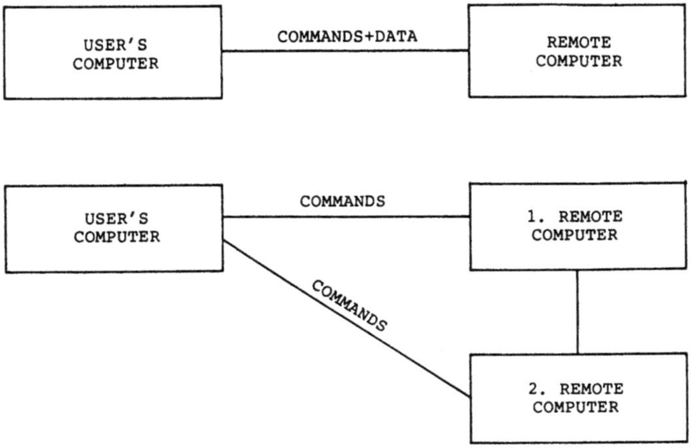

Fig. 2.15. File manipulation

The *Virtual Terminal Service* is based on the concept of the Virtual Terminal. The Virtual Terminal is an abstract object with numerous characteristics of real terminals.

Which characteristics have the real terminals?

— display size (if display terminal),

— buffer,

— functional keys,

— commands.

The virtual terminal is a generalized terminal of any terminal class. It is impossible to define a single virtual terminal because of the numerous real terminals. Therefore the ISO defines 3 virtual terminal classes:

— Basic Class,

— Forms Class,

— Graphic Class.

The parts of a virtual terminal are

— a simulated real terminal,

— an adapter to the standard language.

There is necessary to connect an adapter to the concrete terminal realizing the application of a standard language (Fig. 2.16).

The terminal-handler maps every real terminal type into a virtual terminal.

The third big general protocol group is the *Job Transfer* and *Manipulation*. The job enters into a node of the computer network and is executed in an other remote node. The Job Transfer and Manipulation Protocol is responsible for job operation in distributed systems. The job operations are

— document delivery,

— acknowledgement of the arrival of any document,

— announcement of reports.

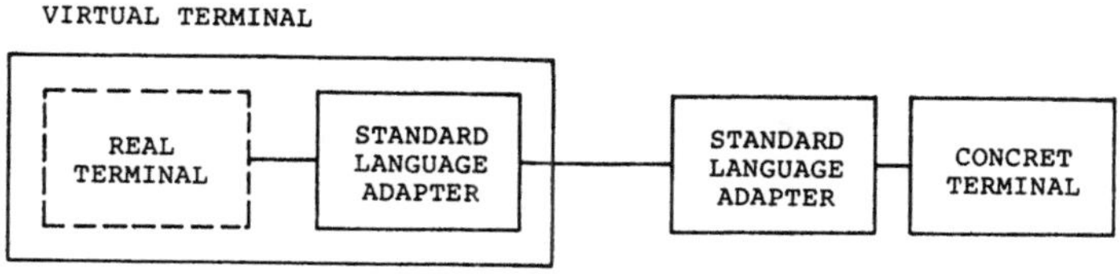

Fig. 2.16. Virtual terminal protocol

Application Layer Structure

The application layer structure can be defined by the categories of the application layer service elements. An application entity composed from service elements can be seen in Fig. 2.17. The service element categories are the following:

— Common Application Service Elements (CASE). These service elements produce for the application process the possibilities which are independent of the nature of the concrete application, e.g. first of all to interconnect the open systems, e.g. login.

— Specific Application Service Elements (SASE). These service elements can be used in a wide range of ways to realize specific information transfer, e.g. data base access.

— User Specific Service Elements (USSE). These provide capabilities to satisfy the needs of particular application processes, e.g. banking.

The partitioning of application service elements into these three categories does not imply the existence of three independent protocols, but corresponds to the groupings of the protocol elements.

In the centre of Fig. 2.17 we can see the user element too. The user element is a part of the application entity and it uses the other service elements.

CASE and SASE create a set of tools for developing application protocols. The elements of CASE are:

— general-purpose service elements,
— connection establishment and release,
— determination of the application context,
— coordination of communication activity.

The elements of SASE are:

— elements for FTAM, VTS and JTM,
— elements for industry-specific application protocols,
— elements for management protocols (system management and application management).

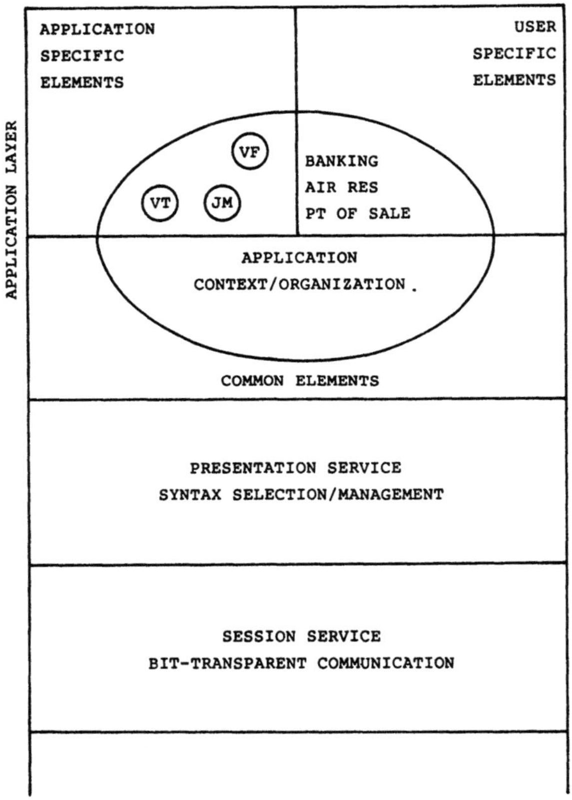

Fig. 2.17. Upper Layer Architecture

3. ABSTRACT MODEL OF TESTING

3.1 PURPOSE OF TESTING

Specification, Verification and Testing

The design of any systems should be begun by specifying the essence of the system's operation. The specification should then be elaborated step-by-step, in an increasingly more detailed way (Fig. 3.1) [MER-74], [MER-79].

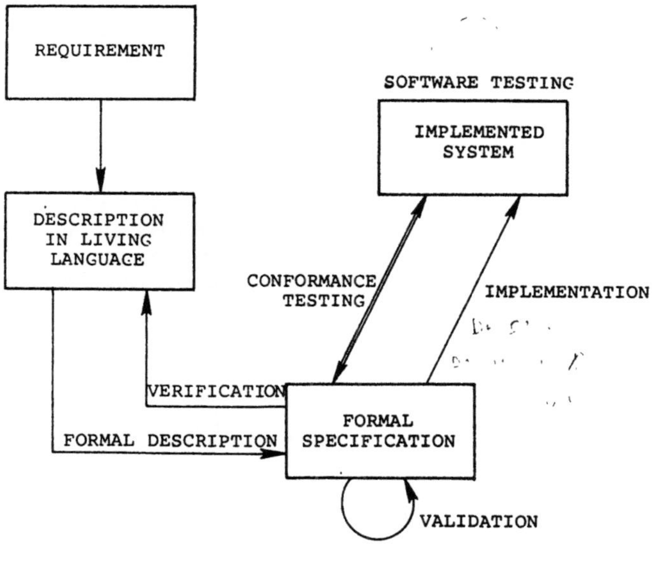

Fig. 3.1. Conformance testing, protocol description and implementation

The dynamic operation rules of the computer networks are described by the protocols, therefore the latter must be specified. Specifying the protocols means to determine the protocol definition built upon its formal description model [BOCH-SUN-80], [KOV-82], [MER-FAR-76]. The protocol definition comprises the syntactic, semantic and temporal rules of the message exchange between the communicating partners in the network. The formal description model comprises the textual description of the protocols, unambiguously and in a way that enables algorithms to be elaborated easily. The protocol specification is the basis for protocol design, testing and verification.

The designed protocol has to be tested to see whether it operates correctly. Some of the conditions for correct operation are:

— the protocol has no deadlock,
— any state can be reached from any other state,
— the protocol returns to its normal operation after suspension of a non-fatal error.

Protocol verification means the checking of the logical correctness and proper operation of the protocol. The verification is less general and more tightly connected to protocol design than the specification. Verification is not discussed in the following parts.

Protocol testing means the control of the protocol that is already in operation. It has two types:

— conformance (identity) testing. This tests whether the characteristics of the implemented protocol conform with those given in protocol specification
— performance measurement. This determines how fast and how reliably can information transfer be provided by the protocol in operation, in the case of different loads [RAY-83], [RAY-86].

Conformance Testing

The national and international standards organizations (ISO, CCITT, ANSI, NBS, etc.) and the bigger computer firms (IBM, DEC, BURROUGHS, etc.) are elaborating their network systems and protocol standards. All of them promise that the modes comprising standard protocols can be connected, without any special effort, to an already operating network. The precondition of this connection is that the implemented protocol should really be equal with the standard one.

The methods of conformance testing have been prepared in order to test the identity.

Why is conformance testing difficult?

— because of the dialogue character of the protocols (a finite correct answer is possible),
— it is difficult to express the test results in numbers,
— the protocols are very complex.

Conformance testing begins by checking the basic and simple things then, advancing step-by-step examines the more complex processes. Testings are performed by layers separately, however the protocols under the layers to be tested should operate correctly. The test of the protocol is made easy by the functional partitioning therefore the test procedures should be elaborated in such a way that they are able to test the protocol functions one-by-one.

Conformance testing is always preceded by the protocol specification. Based on the specification, test sequences can be generated. The test sequence is the central conception of the conformance testing.

What are the requirements for the test results?
— it should be repeatable. A test repeated under the same conditions should give the same results,
— it should be comparable. The results obtained by different testing systems should be comparable.

Performance Measurement

The consequence and after-effect of successful conformance testing is the entry of a new user into the network. If the user has already entered the network, the question is what the network can afford the user:
— what is the performance of the network,
— how robust is it?

The measures of the performance are the different message delays, round trip delays and throughput.

Throughput means the quantity of information transmitted during a time unit. The measure of robustness tells us how and in how many steps the system can be restored from an erroneous situation.

Performance depends very much on load, therefore the delay or round trip delay should be given together with the throughput belonging to it.

A saturated network is better charcterized by throughput, whereas an unsaturated one by the different delays. The business of several resources (channels, nodes, host computers, etc.) belong also to the performance characteristics.

3.2 NETWORK ENVIRONMENT

Characteristics of the Network

In our examination up to now the communication between two picked up nodes was analysed. How is the communication between two entities affected by the other nodes? Let us pose a question, e.g. at the level of data communication, in the following way: How does the traffic between two nodes depend upon the traffic of the other nodes, the characteristics of the nodes and the channels? shortly: on the network environment? The network environment is determined by the topology and the elements of the network.

The topology of the network is described by means of different topological matrices. The general form of these matrices is the following:

$$T = \begin{bmatrix} a_{11} & a_{12} & \cdots & a_{1n} \\ a_{21} & a_{22} & \cdots & a_{2n} \\ \vdots & & & \\ a_{n1} & a_{n2} & \cdots & a_{nn} \end{bmatrix}$$

The number of the rows and columns of the matrix are the same, generally one node belongs to one row and/or column, respectively.

Table 3.1 gives a survey on the most often used topological matrices.

Table 3.1. Topological matrices

Matrix name	Matrix elements
Connection - channel number matrix	Number of channels connecting the nodes
Speed matrix	Channel speed
Cost matrix	Cost per channel
Capacity matrix	Amount of communication paths crossing a channel

The topological matrices determine the relationships among the basic elements of the network, among the nodes and channels.

In order to characterize the network environment the data of the nodes and channels should also be given. The data of the nodes are:

— the number of the node buffers (the quantity of information that can be stored in the node) given in message [packet] or bits,
— the operation speed of the node computers [s],
— the frame processing time in the node computers, by frame types [s],
— the number of host computers connected to the nodes
— the rate of the host channels [bit/s],
— the error probability of the nodes.

Data of the channels:

— the rate of the channel [bit/s],
— the load capacity of the channel (the max. number of information paths crossing through the channel),
— the bit-error rate of the channel (average number of bits transmitted error-free between two erronous bits).

Queues in the Network

The elements of the network, the channels and the nodes are finite resources. The nodes consist of layers corresponding to the rules of the Open System Architecture; the processing time of the layers is also finite.

Where should the messages be waiting in the network?

Everywhere where there are finite resources. The finite resources can be described by means of series and queueing systems. A queueing system can be seen in Fig.3.2.

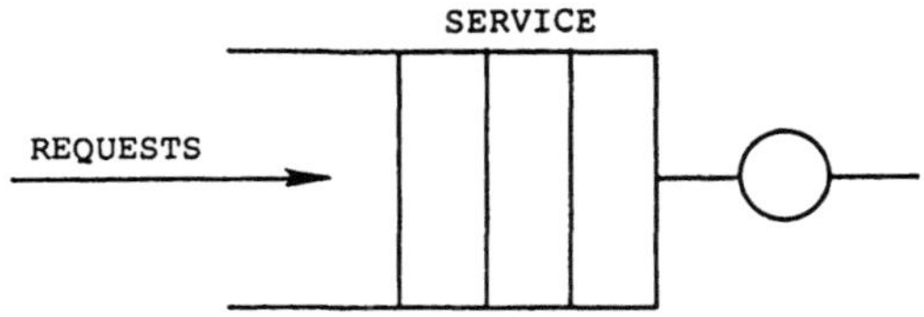

Fig. 3.2. Simple queueing system

The requests are characterized by their arrival intensity and by the statistical fluctuation of the intensity. The arrival intensity means the number of arriving requests during the time unit.

The ratio of the service, i.e. the service intensity means the number of requests processed during the time unit. The service intensity is called system capacity.

The measures of the queueing system are:

— the waiting time of the requests,
— the number of requests waiting in the queueing system,
— the length of the business interval (i.e. the continuous time duration in which the service unit is always busy),
— the duration of idle running time,
— the momentaneous work remainder.

The measures of the queueing system are probability variables, thus their complete characterization is given by their distribution function. In practice, we are generally content with the knowledge of the expected value and the spread, i.e. the first few moments.

The waiting times of the information crossing in the network, can be characterized by queueing systems in the channels, nodes and by the layer interfaces of the open systems.

The Traffic

The traffic is characterized by the length of the arriving messages and the time duration elapsing between the arrivals of the messages. Therefore the main components of the traffic are:

— the distribution of the packet load,
— the time distribution of the packets.

Both components can be deterministic and stochastic.

The deterministic traffic is described by a known and steady arrival intensity. It is easy to avoid congestion if the arrival intensity can be set, in a proper extent, to a smaller value than the capacity of the system.

In the case of stochastic or random traffic the arrival time of the requests is uncertain and cannot be indicated in advance, or the length of the requests also varies by a random character.

The possible components of the traffic are enumerated in Table 3.2. D refers to the deterministic, S to the stochastic character.

Table 3.2. Deterministic and stochastic components of the traffic

number / component	1	2	3	4
Load distribution of the channels	D	D	S	S
Time distribution of the channels	D	S	D	S

3.3 ERROR GENERATORS

Relationship Between the Network Topology and the Errors

The effect of the errors in the network depends on the network topology and on the layered structure. In this chapter we analyse the relationship with the topology, and in the next one we deal with the relationship with the layered structure [PAY-LI-78].

The computer network is — statically — a net in graph theory, i.e. a connected, directed graph without loop-arcs. Any two different places are connected by one path at least in a connected graph and the loop-arc means an edge fitted to a single place.

What graph parameters can be used to describe the network error caused by a graph-element error?

— degree,
— connectivity,
— cohesion.

The *degree* of a node means the number of lines, or arcs coupled to the node. The degree of the graph (δ) is the smallest node degree. The measure of *connectivity* (κ) is how many nodes should be eliminated to disjoin the graph. k-connectivity means that the number of nodes to be eliminated is k. The set of nodes to be eliminated is called node cutset. The connectivity of the graph is the value of the smallest node cutset.

The *cohesion* or line connectivity (λ) is the number of the lines to be eliminated to disjoin the graph. The set of lines to be eliminated is called line cutset. The cohesion of the graph is the value of the smallest line cutset.

The relationship between the above-mentioned graph-parameters is described by the inequality

$$0 \leq \varkappa(G) \leq \lambda(G) \leq \delta(G) \leq \frac{2q}{p}$$

where

$\kappa(G)$ is the graph connectivity
$\lambda(G)$ is the cohesion
$\delta(G)$ is the degree of the graph
q is the number of lines
p is the number of nodes.

From the inequality it is obvious that the number of lines and nodes determines the upper limits to the graph parameters.

The reliability of the network depends on the proportion of the not-communicating node-pairs because of errors and because of the maximum line number

$$R = 1 - \frac{N(P_1, P_n)}{q_{max}}$$

where

$N(P_1, P_n)$ is the expected value of not-communicating node-pairs
q_{max} is the maximum number of the lines.

The expected value of the number of the not-communicating node-pairs depends on the node errors and line errors

$$N(P_1, P_n) = \sum_{i=1}^{P} n(i) x_i + \sum_{i=1}^{q} m(i) y_i$$

where

$n(i)$ is the expected value of the error in the node i
$m(i)$ is the expected value of the error in the i-th line
x_i and y_i probability distributions.

There is often an adequate approximation to assume a binomial distribution, in that case

$$x_i = \binom{P}{i} P_n^{i} 1 - P_n^{p-i}$$

$$y_i = \binom{q}{i} P_1^{i} 1 - P_1^{q-i}$$

E x a m p l e 3.1. Compute the characteristics of the network (Fig. 3.3) given below.

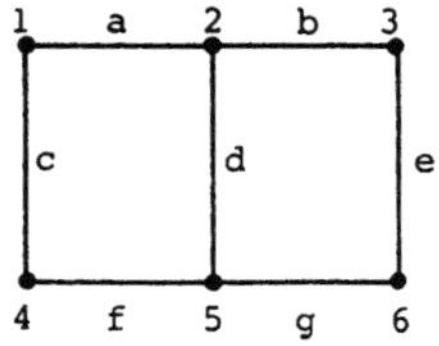

Fig. 3.3. A network (Example 3.1)

The degree of nodes 1, 3, 4 and 6 is 2; the degree of nodes 2 and 5 is 3. Therefore the degree of the graph is

$$\delta(G) = 2$$

The figure shows that a minimum of 2 nodes (viz. 2 and 5) must be removed to divide the graph into two parts, hence

$$\kappa(G) = 2$$

The cohesion or line connectivity can be brought back to the elimination of a and f, or b and g, or c, d, e. Therefore the minimal line cutset is

$$\lambda(G) = 2$$

The inequality condition between the network parameters is fulfilled:

$$0 \leq 2 \leq 2 \leq 2 \leq \frac{14}{6}$$

Assume that no node can get out of order (e.g. a reserve node switch on in the case of an error) and the expected value of line error should be the same in each line

$$P_n = 0$$

$$P_l = 10^{-6}$$

Therefore

$$N(P_l, P_n) = \sum_{i=1}^{7} 1*10^{-6} = 7*10^{-6}$$

The reliability of the network is

$$R = 1 - \frac{7*10^{-6}}{7} = 1-10^{-6}$$

E x a m p l e 3.2. We should connect 6 towns with each other to get a better network reliability than 0.95.

A five-kind connection of the 6 node can be seen in Fig. 3.4, the reliability curves in Fig. 3.5.

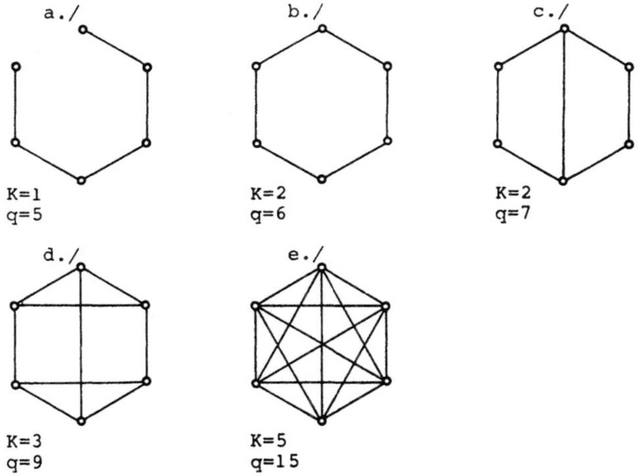

Fig. 3.4. Five possible connections of the six towns

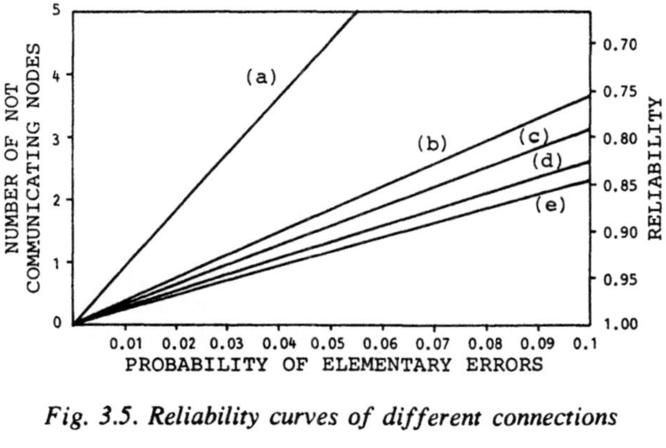

Fig. 3.5. Reliability curves of different connections

We can read from the reliability curves that the elementary error probability is 0.03 by fully connected net and 0.005 by chain. The probabilities belonging to other topologies are between these two extreme values.

Relationship Between the Layered Structure and the Errors

The Reference Model is basically composed of two parts:

— data communication and

— data processing.

The first of the measures for communication errors is bit- and block-error rate.The bit-error rate (BER) is the average number of error-free transmitted bits between two erroneous bits.Then the value of the line error is an exponential distribution supposing

$$P_1 = 1-e^{-\frac{L}{BER}}$$

where

L is the length of the frame bit.

The block-error rate (BLER) is more characteristic on the errors because of transient effects and the bursty noise.

The Reference Model has in every layer two kinds of message length:

— the length of the information message,

— the length of the control message.

Depending on these lengths the appropriate error-values can be gained from two distributions.

The basis of error recognition is the acknowledgement. Every item of information is individual. The various items of information can be differentiated from each other by different sequence numbers. The acknowledgement refers to the sequence number and therefore the control is simple [BU-82].

What kind of error can be detected by the acknowledgement?

— a data unit is lost

— a data unit arrives twice

— the segmenting of the data units is in erroneous sequence

— the blocking of the data units is in erroneous sequence.

The sender repeats the data units; it sends them again if the acknowledgement does not arrive within the time-out.

The resetting can also help in an erroneous situation. The resetting means the whole clearing of the connection from the data units and the restarting of the flow control. Though the resetting helps in erroneous connections, the resetting itself may cause new errors.

Error Generators of the Protocol Testers

It is not enough to know that the system operates correctly under normal conditions, the events originating from erroneous operation should be tested, too. The renewed testing detects the appearance of the error and not its absence, therefore erroneous situation should be produced and the reactions of the implemented protocol should be analysed. The error generators perform this task.

The objectives of the error generators are:

— to generate syntactically correct data unit sequences in incorrect context (the surrounding text is erroneous),
— to generate syntactically erroneous protocol data units,
— to interpret the syntactically incorrect protocol data units by reception,
— to alter the predetermined time-outs,
— to generate semantically erroneous responses,
— to generate errors in the adjacent layers.

The faultless, correct reference protocol can neither send nor receive sequences being erroneous syntactically, semantically, or temporally. This is the task of the error generators.

Figures 3.6-3.8 demonstrate three methods how the error generators can be connected to the active tester.

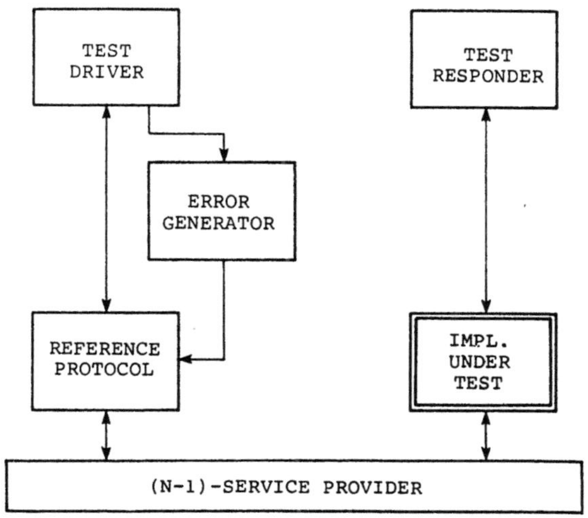

Fig. 3.6. Error generator controlling the parameters of the reference protocol

Test driver and test responder are needed in all three methods.

The error generator controls the internal variables of the reference protocol in Fig. 3.6. In this manner the protocol behaviour can be changed.

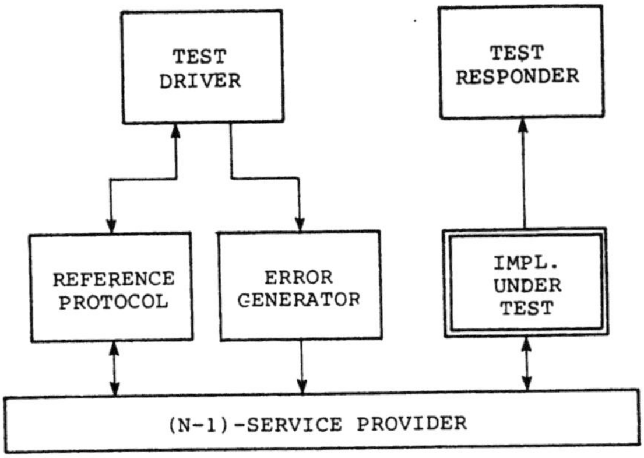

Fig. 3.7. Independent error generator

The error generator in Fig. 3.7 produces, directly, protocol data units inserted any comprehensive information sequence. There are essentially more possibilities f changing, but the coordination of the error generator and reference protocol caus many problems which are difficult to solve.

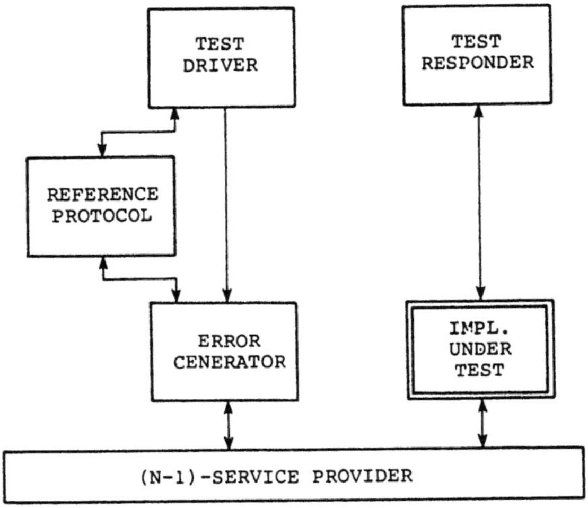

Fig. 3.8. Strongly connected error generator

Error generation according to Fig. 3.8 seems to be the best solution. It is not necessary to coordinate the two units. Syntactic, semantic and time-out errors can be produced very simply and the appropriate input for the reference protocol can be ensured by decoding by the reception. There is the possibility for creating an erroneous data environment to error-free data units.

3.4 CONFORMANCE TESTING

Requirements

The purpose of the OSI is to let interwork heterogeneous systems be implemented in a different way and independently from each other. The assumption of interworking is the existing protocol according to the standards. The requirements of conformance testing must be worked out to justify the identity of the implemented protocol and the standard unambiguously.

When does a protocol fulfil the requirements? What kind of requirements should be realized and which realization should be tested?

The requirements can be classified according to various points of view:

— the type of requirements,

— on which protocol properties dö the requirements refer?

The classification according to the type of the requirements:

— mandatory requirements
  these should be tested in any case e.g. the test of flags in HDLC-frames, because every frame begins and ends with a flag,

— conditional requirements
  these requirements are referred to properties which are true after fulfilling some predetermined conditions
  e.g. the control field of HDLC-frames is composed of 16 bits, if it is extended mode (modulo 128),

— prohibitions
  these are states and events which do not occur in normal case
  e.g. a slave cannot send a message in the NRM mode (HDLC) until the master sends permission with a P poll bit,

— optional requirements
  these concern the properties which are free to choose
  e.g. the t1 time-out in HDLS depends on an agreement of the Postal Administrations.

The optional requirements can be classified further depending on the possible number which can be chosen from and on the type of the choice (whether it is mandatory or not).

If the choice is of Boole type, then it is necessary to choose one from the two possibilities. A mutually exclusive choice means a mandatory choice of one value from two values. The testing standards differ yet what can be chosen (n number from m values) or what is necessary to choose (n number from m values).

The other important question is the requirements to which protocol properties refer.

The classification of the requirements according to the internal properties are the following:

- messages and their parameters,
- time periods,
- states,
- protocol mechanisms.

Messages and their parameters

To test the message (Protocol Data Units PDUs) and their parameters is the most fundamental control and it is from more points of view the simplest check. Before testing it is worthwhile to consider the following questions concerning every message:

- Is the objective of the message clear? When is it necessary to send it and what are the possible answers,
- Is the permitted set of parameters belonging to the message clearly stated? Is the permitted range of values clearly stated for each parameter,
- How does the protocol handle the illegal messages?
- What happens if the parameter is out-of range,
- Have the data units of different protocols a common parameter,
- Is there a relationship among the independent parameters,
- Is it clearly defined in which order the bits, octets should be carried in the underlying protocol,
- Is there an unambiguous relationship between the protocol data units and service data units.

Time periods

- Have explicit upper limits been set for each allowed time period between the first party receiving a message and sending its required response,
- Have separate explicit upper limits been set for the allowed time period that the first party should wait for a message from the second party that does not come,
- Have explicit lower limits been set for such time periods.

States

To define the protocol procedure, finite state machine models are often used. Which questions are then necessary to be answered:

- Are all states clearly defined?
- Is there any event occurring only in a subset of the states?
- Are the required actions and state transitions defined for each possible state-event pair?

— Is it always possible to detect the state of an implementation from observing the inputs and outputs?
— What happens if possible but invalid state-event pairs occur?

Protocol mechanisms

— Are there any restrictions on the number of successive times that a recovery mechanism can be involved?
— If a window mechanism is used, is it fully defined?
— If the protocol contains a negotiation mechanism, is this clearly defined?

After the clarification of the above mentioned questions it is useful to work out a schedule for the conformance testing of the protocol, e.g. the following is one of the possible sequences

— to check the common elements of the messages
  e.g. flag at the beginning and the end of a message format of LAPB,
— unnumbered messages
  e.g. DISC in LAPB,
— numbered messages
  e.g. RR in LAPB,
— messages with adjustable parameters,
— time between the stimulus-message and response-message,
— maximum waiting time,
— handling a message arriving with minimum delay,
— every state transition in the phase of connection establishment and release,
— every state transition in the phase of data transfer,
— sequence numbering,
— window mechanism,
— priorities,
— segmenting,
— multiplexing.

The Classification of Conformance Testing

Conformance testing can be classified according to the point of view of the tester or of the system to be tested.

The tester can be passive making only observations or active generating sequences. The testing can be executed from a remote node or in the node to be tested.

The testing is manual when the operator must write the test commands; it is automatic when the test steps following each other are computer-controlled.

The advanced versions of the active testers include network environment manipulator unit and error generator, too.

The testing can concern either a simple layer or more layers of the system to be tested (single-layer testing, multilayer testing).

The protocol testing can be functionally decomposed:

— connection establishment and release,
— data transfer,
— error handling.

The decomposing of the testing can also follow the character of the functional behaviour:

— normal,
— abnormal,
— illegal.

We have assumed until now that the system to be tested has an open system architecture. If the system to be tested completely differs from the open systems, then the network owner or developer has sole responsibility for the specification of the testing.

We are not dealing with this case but we are interested in systems which partially have OSI structure (Fig. 3.9).

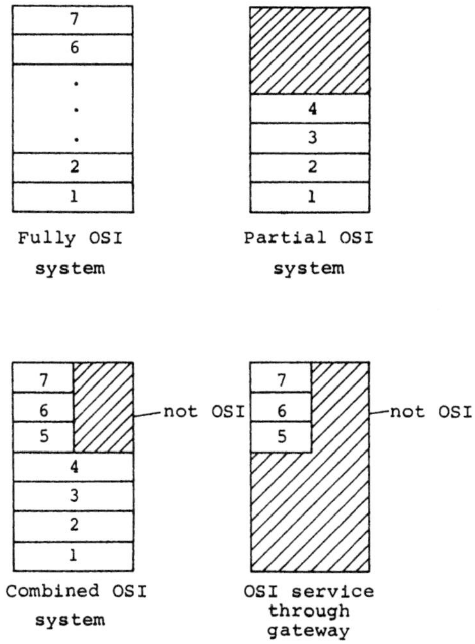

Fig. 3.9. Classification of the systems to be tested comparing to OSI

3.5 THE ABSTRACT MODEL OF CONFORMANCE TESTING

The Fundamental Concept of Abstract Testing

Initially, conformance testing was realized with individual experiments, the methods were difficult to generate and difficult to repeat after changing the network environment. The solution was the composition of the abstract testing method similarly to the Reference Model of the open systems.

The abstract model of conformance testing is based on two fundamental terms, these are,

— the controllability,

— the observability.

The protocol entity is a black box, only the input and output are available. Protocol data units and service primitives can be controlled and observed at the input and output (Fig. 3.10).

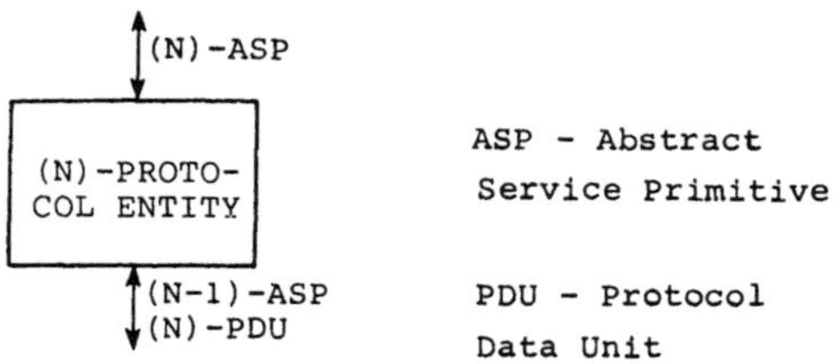

Fig. 3.10. The protocol entity

The identity of the protocol to be tested and the reference protocol entity can be measured on the surface fitting to the lower or upper layer of the protocol entity [GAST-GU-85], [CAST-86].

The testing can be executed from a remote node, from a test centre or in the node to be tested, but test functions can also be distributed between the remote node and the examined node.

The testing possibilities are somewhat generally demonstrated in Fig. 3.11. The controllable and observable protocol data units and abstract service primitives are introduced in the figure. The (N)-reference entity indicates the basis of the comparison, the standard protocol entity. The notation " distinguishes the remote data units and service primitives from the units and primitives in the node to be examined.

There are four abstract testing methods suggested in ISO:

— Local test methods. These use control and observation of the ASPs directly above and below the Implementation Under Test (IUT);

— Distributed test methods. These use control and observation of the (N-1)-ASP"s together with control and observation directly above the IUT;

— Coordinated test methods. A test management protocol controls the upper tester;
— Remote test methods. These use control and observation of the (N-1)-ASP"s, with the ASP activity directly above the IUT hidden.

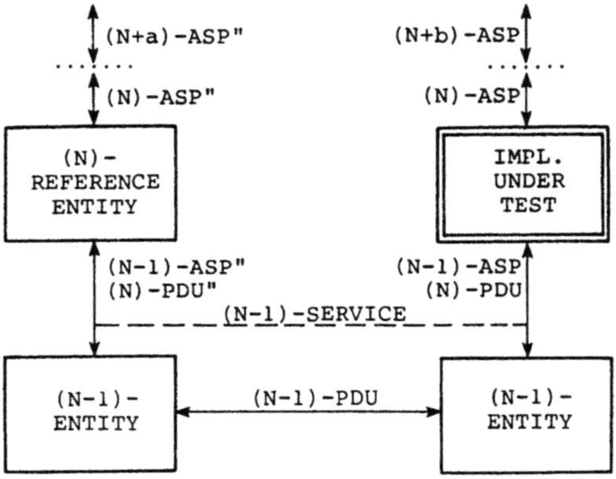

Fig. 3.11. Conformance testing

Methods of Conformance Testing

The *local test method* means the control and observation of abstract service primitives in the layers that are above and below the protocol entity to be tested (Fig. 3.12).

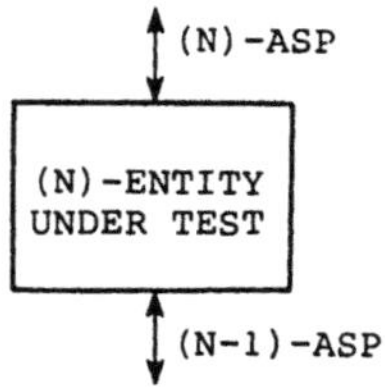

Fig. 3.12. The local test method

The local test method can be used in two different ways: we can change the service primitives above the IUT and observe the service primitives below the IUT or we can control the lower layer service primitives and observe the upper layer primitives.

Local testing is the fundamental of all methods because this is the control of an isolated basic unit picked up from the environment. This direct testing is useful for the implementor, it is easy to execute and it is easy to evaluate. Direct testing is the basic element of multilayer testing. The disadvantage of local testing is that the testing does not occur in a real communication environment.

Distributed test methods. In this method, the IUT does not have a PCO at the lower service boundary. Therefore, (N)-ASPs and (N)-PDUs are controlled and observed indirectly at the remote end of the (N-1)-Service Provider allowing the upper and lower testers to reside in physically separate locations.

The *remote method* is realized in real remote communication environment. The two functions of testing, viz. control and observation, are separated. The principle of testing is the following (Fig. 3.13).

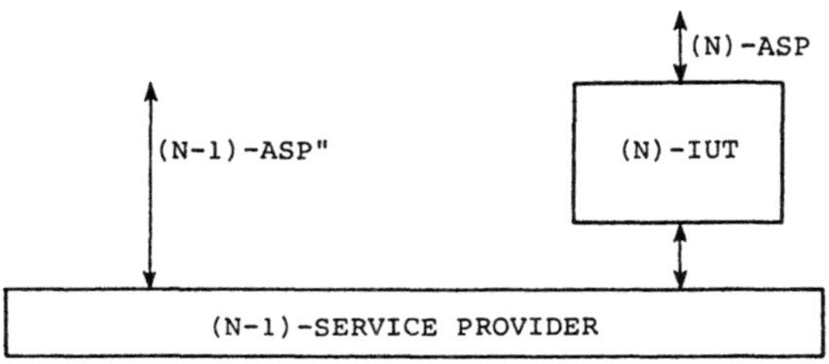

Fig. 3.13. The distributed test method

One realization of the testing is

— the IUT is controlled on the upper interface manually or automatically through a special test interface,

— the service primitives appearing at the lower interface of the IUT are observed.

The other realization of the testing is

— the IUT is controlled from the reference protocol side through the (N-1)-service provider,

— the IUT-answers to the upper layer are observed and evaluated.

The remote and indirect method gives much more truthful results than local testing. Its disadvantage is that some service primitives can be produced with difficulty. The remote method means remote control and remote observation. The partner implementing a new protocol should not build-in special testing tools in his system. It is favourable that the same person at the same place carries out the control and the evaluation of the received answers. The remote and hidden method controls the (N-1)-service primitives àt the reference side and observes the (N-1)-service primitives arriving likewise at the reference side as the responses of the side to be tested (Fig. 3.14).

This method is very advantageous when the IUT, the OSI protocol to be tested is implemented in a front-end processor which maps this protocol into a non-ISO protocol.

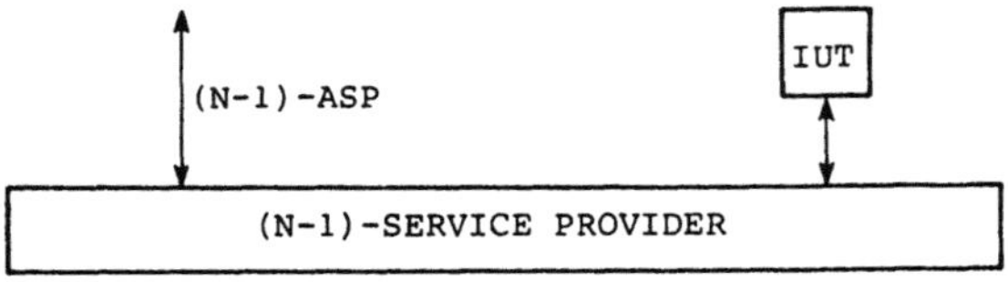

Fig. 3.14. The remote test method

Coordinated conformance testing is a complex and sophisticated method. The lower and upper tester has a special connection realized by a test management protocol. The test events are remote abstract primitives of the lower layer, protocol data units and test management protocol data units.

The test configuration can be seen in Fig. 3.15. This method is an advanced version of the distributed conformance testing. It has two additional features: the coordinating process and the standard test management protocol.

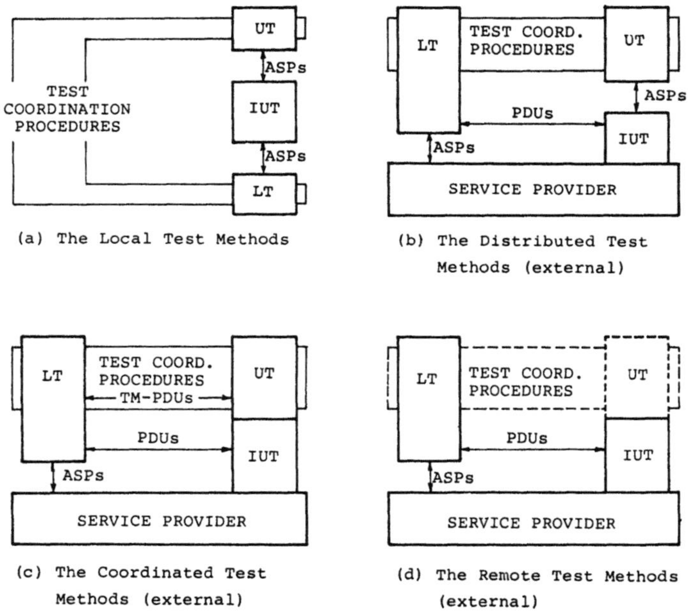

Fig. 3.15. Overview of Abstract Test Methods

The General Logical Architecture of Conformance Testers

The architecture of the conformance testers has a layered structure similarly to the Reference Model. The reference protocol and Implementation Under Test (IUT) are placed in the same layer (Fig. 3.15). The test driver allows the testing to start and the reference protocol stimulates the entity to be tested with the reference service primitives. The test responder registers and evaluates the reactions on the stimuli arriving from the IUT and initiates — when required — new messages from the reference side to the protocol to be tested. The test-driver also makes observations and evaluations.

The conformance testing must clear up the protocol behaviour under erroneous circumstances. It is therefore necessary to complete the reference node with additional units in order

— to generate valid and invalid messages in the same way,
— to receive exceptional messages differing from the normal messages.

The Physical Architecture of Conformance Testers

The test centre is a single active tester in the simplest case. The tester is realized by a mini-computer and the node to be tested is connected to the test centre either through a public data net or through other communication media. A very simple configuration is Fig. 3.16.

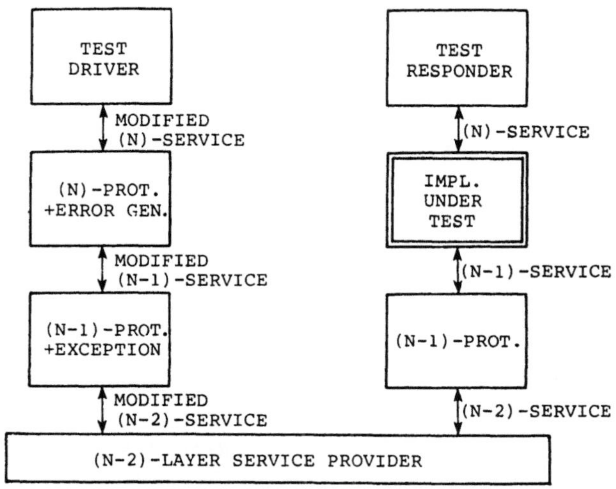

Fig. 3.16. Logical architecture of the conformance testers

This configuration is suitable for the (N)-layer testing, when the (N-1)-connection is an end-to-end connection (alive). If this connection is not fulfilled (e.g. the third level of X.25), a somewhat more complex test system is needed (Fig. 3.17).

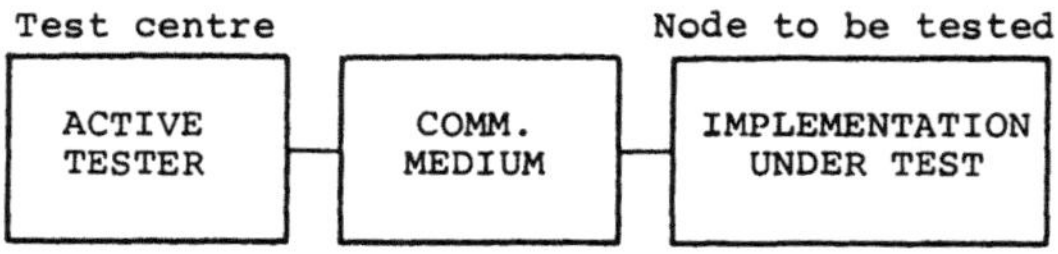

Fig. 3.17. A simple tester

The Active Tester (AT) is composed of two parts:
— test driver,
— Environment Manipulator Unit (EMU).
The environment manipulator unit imitates the network environment and realizes every logical testing function; the test driver ensures the synchronous operation.

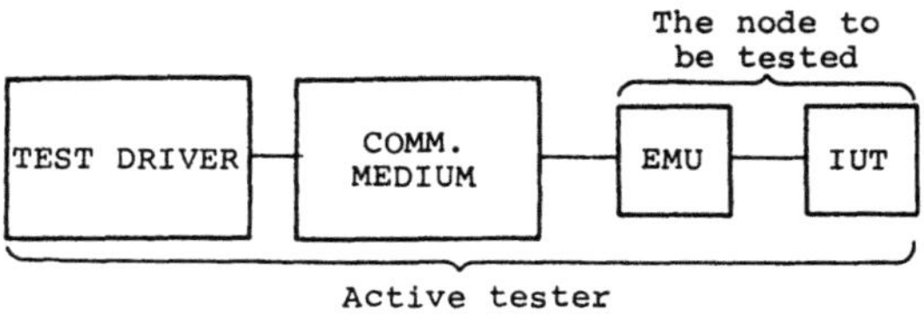

Fig. 3.18. An active tester

An Example on Conformance Testing

The succeeding steps of the conformance testing for LAPB are enumerated below as an example:
— Syntactic check,
- common elements of the frames (flag, address, FCS);
- unnumbered frames: SABM, UA, DISC, FRMR;
- supervisory frames: RR, RNR, REJ, SREJ the control is first compleated by fixed N(R), the N(R) is changed one by one;
- information frame
  first with fixed N(R) and N(S), then N(R) and N(S) alternating changed one by one, finally both are changed simultaneously.

If an error is found in the syntactic check, the testing should be continued only after error-correction.

— Functional control
  - first the side to be tested is the receiver;
    how it receives the intention of the other side to establish a connection, to transfer data and to release;
    does the acknowledgement mechanism operate correctly?
    are the time-out values correct?
  - the side to be tested is the sender
    every examination should be repeated.

All message types should occur at least once by the functional control.

— Examination of the error-handling properties
  - how does the IUT receive the syntactically erroneous frames?
  - is the I-frame repeated, when the acknowledge does not arrive within the prescribed time out?
  - are the REJ and SREJ sent correctly?
  - are the REJ and SREJ interpreted correctly?
  - how are the wrong addresses interpreted?
  - what is the IUT reaction on the reception of the bad-sequence-frames?
  - how are the collision of unnumbered commands handled?
  - do the resetting procedure operate well in both directions?

— Test sequences

It we get correct result of all the above-enumerated tests, the examination of the protocol can start with complex test sequences.

The error should be earlier corrected by any semantic or timing error and the test continued only after the correction.

(More methods will be introduced for complex test sequence generation in Part III.)

There are many testing strategies for the LAPB, too. The schedule in our example is only one possible solution.

3.6 PERFORMANCE TESTING

Requirements

The physical definition of power is the work that a force can execute during a given time.

What is the "work" of the computer network? It is: to deliver messages from one node another during a finite time.

The number of transported messages and the transmission time characterize together the performance (the power) of the network. Primarily the time was selected as a performance measure, but the number of the messages, the throughput was allocated to the time as a parameter [ECS-TAR-86], [ECS-TAR-87].

The performance measure can be either the one-way delay which is the time necessary for a message advancing from the source to the destination or the round-trip delay which is the travelling time of a message from the source to the destination and back: from the destination to the source.

Both delays are closely related to the throughput, therefore the delays are generally measured by three types of network load:

— heavy,
— average,
— weak.

The delays are influenced by the numbers of multiplexing. It is useful first to measure the connection without multiplexing, after that the test should be repeated with multiplexing changing the number of multiplexing (if many messages travel between two hosts, there is more congestion and more waiting).

The layered structure of the nodes influences the network performance, too, and therefore the delays are correctly interpreted between the source and destination of the same layers.

The Schedule of Performance Testing

Conformance testing precedes performance testing. If our system has identical syntactic, semantic and time-out features with the predetermined, then the robustness test follows. The essence of the robustness test is to determine how many steps are needed to recover from an erroneous situation.

The effect of the following errors should be controlled:

— wrong user-commands,
— protocol error,
— communication error,
— hardware system error,
— software error.

The performance test system can cause errors, too, if the rate of the channel connecting the tester host to the tested host is in the order of the network part to be measured. Therefore a high-speed channel is necessary between the tester host and the connecting network host.

After the conformance testing and error handling control, the performance testing can be executed in the under-mentioned sequence:

a) Delay among the data link protocol entities

- the network load is
  — weak
  — average
  — heavy

  and no multiplexing
- the network load is
  — weak
  — average
  — heavy

  and multiplexing

b) Delay among the transport layer entities
(loading and multiplexing as above)

c) Delay among the application layer entities
(loading and multiplexing as above)

Naturally the complete, thorough performance testing contains the control of all the seven layers, the suggested performance testing of the selected three layers is the minimum test.

The parameters of the resulted delays are
— the name of the layer and its characteristics,
— the network load,
— the throughput,
— the measure of multiplexing,
— the data of the tester.

Parameters of Performance Testing in the OSI Model

The quality of the layer services of the OSI model can be characterized by various performance parameters. It is useful that the performance measures of the layer services also follow the modularity and flexibility of the open system architecture, therefore the performance measures are referred to two Reference Model characteristics: to the connection and to the service access point [BAR-84].

These are the connection-oriented parameters:
— connection establishment delay,
— normal data transfer delay,
— expedited data transfer delay,
— disconnection delay,
— error parameter,
— throughput.

The service-access-point-oriented parameters are:
— throughput,
— connectivity,
— life-time.

The connection-oriented parameters relate either to a single i-connection or to more, multiplex-connection. For simplicity we enumerate now the definition of the parameters belonging to a single i-connection.

The connection establishment delay is the time necessary to establish a connection:

$$\tau = t_{CON\text{-}conf\text{-}i} - t_{CON\text{-}req\text{-}i}$$

where

$t_{CON\text{-}conf\text{-}i}$ is the point of time when the user of the i-connection received the CON-conf primitive

$t_{CON\text{-}req\text{-}i}$ is the point of time when the user of the i-connection sent the CON-req primitive

The normal data transfer delay between the appropriate layer entities means the delay of data transported through the i-connection

$$\tau_{DT2} = \frac{1}{n}\sum_{j=1}^{n}\left[t_{DT-ind-i}(j) - t_{DT-req-i}(j)\right]$$

where

- n is the number of service data units (SDU) transported during the observation
- $T_{DT\text{-}ind\text{-}i}$ (j) is the point of time when the user of the i-connection received the DATA-ind primitive detecting the start of the j-th SDU
- $t_{DT\text{-}req\text{-}i}$ (j) is the point of time when the user of the i-connection sent the DATA-req primitive detecting the start of the j-th SDU.

The expedited data transfer delay is also an average value similarly to the above, but the expedited data units are replaced in the place of normal data.

The end of the connection is indicated with the disconnection delay:

$$\tau_{Ti} = t_{DISC-ind-i} - t_{DISC-req-i}$$

where

- $t_{DISC\text{-}req/i}$ is the point of time when the user of the i-connection received the DISC-req service primitive
- $t_{DISC\text{-}ind\text{-}i}$ is the point of time when the user of the i-connection sent the DISC-ind service primitive

The error parameter is expressed as the rate of erroneous protocol data units (PDU) to all PDU:

$$U = \frac{n_e}{n}$$

where

- n_e is the number of erroneous PDU-s caused by message duplication, message lost, incorrect addressing, etc.
- n is the number of all PDU-s during the observation.

The throughput characterizes the amount of information crossing the i-connection:

$$T_{Ci} = \frac{\beta_i}{t}$$

where

- β_i is the number of information bits passing the i-connection during the observation
- t the observation interval

The second group of performance characteristics concerns the service access point. The throughput was earlier referred to a single i-connection. The maximal connection throughput is the maximum amount of information transported through a single i-connection assuming that all other connections are free.

The throughput related to the service access point concerns all connections crossing the access point:

$$\max \sum_{i=1}^{N} T_{Ci}$$

where

- N is the number of active connections crossing the access point
- T_{Ci} is the throughput of the i-connection

The connectivity characterizes the ability of the service provider as to how the provider can establish the connections required by the user:

$$C = \frac{n_a - n_b - n_c}{n_a - n_b}$$

where

- n_a is the complete number of connection establishment requests in the service access point
- n_b is the number of unsuccessful connection requests caused by the user (e.g. incorrect addressing)
- n_c is the number of unsuccessful requests (e.g. because of congestion, temporary error)

The average connection life time is the rate of disconnection requests reported by the service provider during the observation

$$\bar{\tau} = \int_0^{\infty} \left[1 - \int_0^{t} d_\tau \, d\tau\right] dt$$

where

- d_τ is the rate of disconnection requests during the connection.

PART II
FORMAL DESCRIPTION TECHNIQUES

4. STATE-TRANSITION BASED MODELS

4.1 OVERVIEW OF FINITE-STATE MACHINES

First a natural language, generally English, was used to describe the architecture of computer networks and to formulate the conditions how a new partner can enter in a network. However words and sentences are ambiguous, have more than one meaning (see, e.g. footnote on p. 117). English people interpret sometimes an expression in a quite another way than others. This often leads to misunderstandings. This is especially annoying when constructing an international computer network. There is another difficulty: the system specification in a natural language is very far from the description needed for implementation. This gap should be reduced. The widespread use of computer networks requires more and more alternatives and new implementations, in view of which an unambiguous formal description preparing the data processing is most definitely necessary.

What requirements must be met by formal description in the field of computer networks, protocols and open systems? [VI-TE-BO-83]. The requirements are:

— to formulate the specification concisely,
— it should be easy-to-learn and easy-to-use,
— it should be well readable,
— an unambiguous, precise and consistent description should be realized,
— avoid overspecification,
— the description should concentrate on the normal cases and not on the extreme and rare events,
— the modularity, decomposition and composition should be supported,
— the automatization and data processing of the specification should be prepared.

The abstraction level of the formal description extends from the network architecture to the dialogue realized by a protocol. A good and appropriate formal description is easy-to-handle and has general validity.

The formal description is the basis of the protocol design, implementation, verification and testing.

Numerous standardization organisations (ISO, CCITT, NBS, etc.) developed concrete suggestions for the realization of formal descriptions. The majority of the formal descriptions model the network operation rules with finite-state machines.

Basic Set Theory

We review in this section the basic concepts of the set theory used for formal description and finite-state machines. A *set* is the totality of objects which can be distinguished from each other based on any common characteristics. A set can be given either by the enumeration of the elements

$$H = \{A, B, C\}$$

or by a general element and a statement describing the set elements

$$H = \{x \mid P(x)\}$$

P(x) is the statement of the elements of the set and of which the set is composed of, the elements x which P(x) is true of.

Some often used notation is the following:

Notation	Meaning
$A \in H$	A is the element of H
$A \notin H$	A is not the element of H
$A \subset H$	A is a subset of H
$\emptyset$	empty set
$A \cup B$	the union of A and B
$A \cap B$	the intersection of A and B

There are mainly string sets in the formal description of computer networks. The elements of string sets are symbols or strings. A finite, non-empty set of symbols is named alphabet. A string is the finite sequence of symbols derived from the alphabet.

Every string x is characterized by its length l(x) .

The length of an empty string is zero

$$l(\varepsilon) = 0$$

Two operations, viz. the concatenation and the power, are used above all in string sets. The concatenation of the elements of two sets is the ordered sequence of the elements of two string sets

set x set x

The product of string sets is the set of concatenated strings. The first component of the concatenation belongs to the first set, the second component to the second set.

$$X_1 \quad Y_1 \quad X_2 \quad Y_2 \quad \ldots \quad X_n \quad Y_n$$

The string power means the self-concatenation of the string

SET X   SET X

The mapping or function determines the relationship between two sets. The mapping is denoted

$$f : S \rightarrow T$$

A mapping f is a relationship which allocates one component of set T to each component of the set S. A function can be defined simply by enumeration of the elements allocated to each other by the function as well-ordered pairs

$$f = \{(s_1,t_1), (s_2,t_2)...\}$$

where

$s_i \in S$
$t_i \in T$ and $i = 1, 2, ...$

Computability. Turing Machines

The claim of a general computing tool was laid more and more intensively in the thirties on the computation of every *computable* function. The *Turing machine* was created as a solution composed of a finite-state machine, an infinite tape and a writer-reader header (Fig. 4.1). The tape contains symbols, the machine is characterized by its states. The operation of the Turing machine consists of the following steps:

— reads a symbol,
— the machine changes its state,
— writes over the symbol on the type that has been read,
— the writer-reader header steps one right or left,
— it reads a symbol again

.
.
.

The advantage of this model is that the computing of the procedure is decomposed into steps following each other and only one change can take place at a time (the reading or writing of a symbol, a state transition of the machine or one step of the writer-reader header). The algorithm is a series of automatically executable steps which terminates during a finite time. Therefore we can rightly suppose that an equivalent Turing machine can correspond to every exactly defined algorithm.

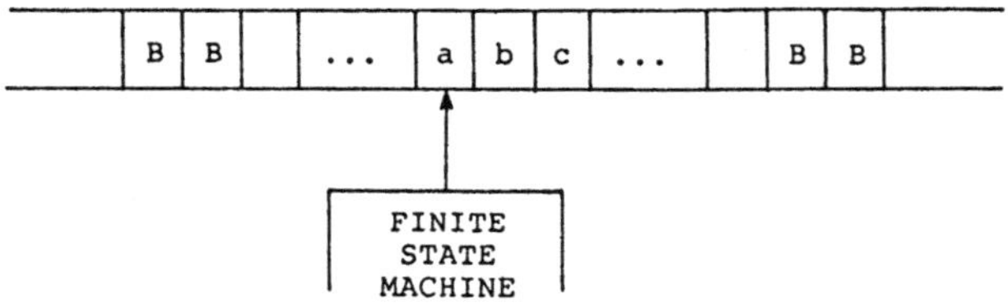

Fig. 4.1. A simple Turing machine

Finite-state Machines

A finite-state machine is an important component of the Turing machine, their relationship can be formulated in this way (Fig. 4.2):

— the finite-state machine is the simplest tool for computing,

— the Turing machine is the most general tool for computing.

The finite-state machine has a number of internal states and it can interpret an input symbol set, an input alphabet. An output symbol appears on the output of the machine produced on the effect of a determined input depending on the momentary state of the machine. Turing introduced the concept of *state*; Shannon, Huffmann, Kleene and Moore applied it in their fundamental works of computer theory and information theory. The intuitive definition of the state is:

The state contains all information characterizing the history of the system and enabling it to be computed in the future. The history of the past can influence the future only within finite steps.

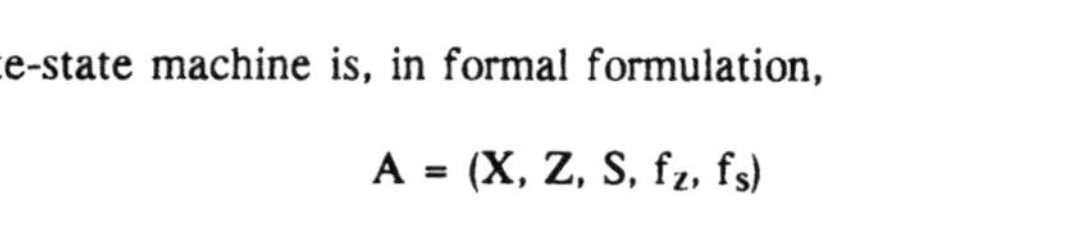

Fig. 4.2. Finite-state machine

The finite-state machine is, in formal formulation,

$$A = (X, Z, S, f_z, f_s)$$

In this 5-tuple

$X = \{x_1, x_2, \dots x_p\}$ is a finite input alphabet

$Z = \{z_1, z_2, \dots z_q\}$ is a finite output alphabet

$S = \{S_1, S_2, \dots s_n\}$ is a finite set of states

f_z is the output function
f_s is the state transition or next state function

The f_z and f_s mappings can be determined from the relationships

$$z_\nu = f_z(x_\nu, s_\nu)$$

$$s_{\nu+1} = f_s(x_\nu, s_\nu)$$

The first mapping corresponds to an output symbol of the momentary input symbol and actual state, the second mapping, however, corresponds to the next state to them.

State-transition Tables and Graphs

If we know the input and output alphabet, the state set and the two mapping function of a given system, the system description can be formalized as a table, as a graph, or as a matrix. The table, graph and matrix are alternative formats of the mappings.

There are many types of the *state-transition tables*, we will use the format according to Table 4.1. The elements of the state set belong to the rows and the input alphabet symbols to the columns. To each square of the table the next state and the convenient output symbol are described.

Table 4.1. State-transition table

Input / State	x_1	x_2	x_3	...
s_1	...	...	...	
s_2	⋮	s_{next}/z_m	⋮	
⋮	⋮		⋮	

The state transitions can be represented by *state-transition graphs* (Fig. 4.3). The state-transition graph is a directed graph in which nodes correspond to the different states of the machine and their arcs mean the state transitions according to the effect of the certain input signals.

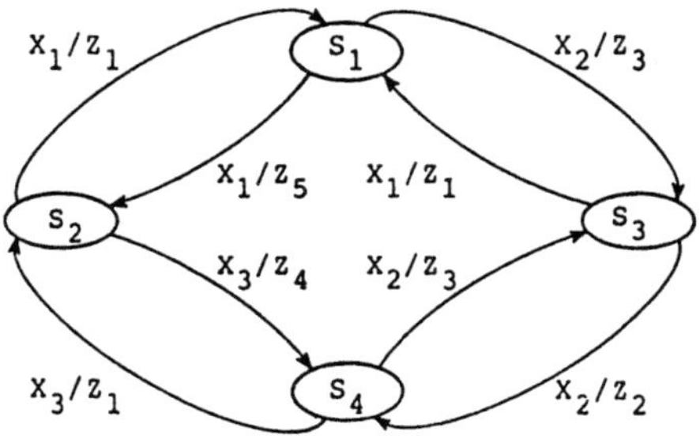

$X = \{x_1, x_2, x_3\}$
$Z = \{z_1, z_2, z_3, z_4, z_5\}$
$S = \{s_1, s_2, s_3, s_4\}$

Fig. 4.3. State-transition graph

Table 4.2 is the equivalent state-transition table of the state-transition graph in Fig. 4.3 [GILL-62], [GILL-63].

Table 4.2. State-transition table for Fig. 4.3

State \ Input	x_1	x_2	x_3
s_1	$s2/z5$	$s3/z3$	-
s_2	$s1/z1$	-	$s4/z4$
s_3	$s1/z1$	$s4/z2$	-
s_4	-	$s3/z3$	$s2/z1$

EXAMPLES

Example 4.1

A wheel turns clockwise and counter-clockwise. A switch is switched alternatively to right and to left.

If the switch is switched in the opposite direction of the turn of the wheel, then the wheel changes its turn direction and a lamp lights up. If the switch is switched in the same direction with the turn of the wheel, then the wheel turns further without reversal and the lamp does not light up.

Here the input corresponds to the position of the switch

$X = \{\text{Right, Left}\} = \{X_1, X_2\}$

The states of the wheel are

S = {to turn clockwise,
to turn counterclockwise} =
= $\{S_1, S_2\}$

The output symbols are

Z = {the lamp lights up, the lamp doesn't light up} =
= $\{Z_1, Z_2\}$

The state-transition table is

STATE \ INPUT	X_1	X_2
S_1	S_1/Z_2	S_2/Z_1
S_2	S_1/Z_1	S_2/Z_2

Example 4.2

To construct the state-transition table of a serial adder.
The input alphabet is

$X = \{00, 01, 11, 10\}$

The serial adder has two states:

S = {no carry,
there is carry}
= {N, Y}

The output alphabet is

$Z = \{0, 1\}$

The state-transition table is, considering the above introduced variables,

S \ X	00	01	11	10
N	N/1	N/1	V/0	N/1
V	N/1	V/0	V/1	V/0

4.2 EXAMINATION OF FINITE-STATE MACHINES

The Properties of Finite-state Machines

The state-transition table characterizes unambiguously the finite-state machines, when the next state and the output event correspond to each pair of input event and state. These are the *fully defined* machines. Machines that describe practical systems are generally not fully defined because contain an undefined next state or an undefined output, or both.

Every system and process can be drawn in several ways as a finite-state machine. There is a difference between a *minimized* and a redundant format. The redundant format can transform to a simpler one. The minimized format has no identical states and the number of states cannot be decreased by any means.

Another important characteristic of the finite-state machines is the measure of the connection: How can the system change its state from one to another? A FSM is strongly connected if it is possible to achieve any state from any other state. Let us make this definition more precise: The machine M is termed strongly connected, if there exists such an input sequence for each S_i, S_j state pair of the machine M carrying the machine from state S_j. The input sequence is the finite set of the elements of the input alphabet.

Equivalence of Finite-state Machines

The examination of the state equivalence is one of the most important tasks from the point of view of applications. We employ the notation M/σ for the fact that the machine M is in the state σ. The notion of state equivalence can refer either to the states of two machines or to the states of one machine. With two machines: the state σ_1 of M_1 is equivalent to the σ_2 of machine M_2, if M_1/σ_1 and M_2/σ_2 produce the same output sequence on the effect of any input sequence. A single machine M has equivalent σ_1 and σ_2 states, if M/σ_1 and M/σ_2 produce an identical output sequence on the effect of any input sequence. The state equivalence examination in the case of a single machine is the tool of minimization. If σ_1 and σ_2 are not equivalent then the states are called *distinguishable.* The fundamental tools for examining the finite-state machines are the input sequences analysing either the dissimilarity or the equivalence of the states. These are the *distinguishing sequences.*

The state equivalences can be classified according to the length of the input sequences. The state σ_1 of machine M_1 is said to be k-equivalent with the state σ_2 of machine M_2, when M_1/σ_1 and M_2/σ_2 produce the same output sequence on the effect of a k-length input sequence.

The next important concept is the equivalence of two machines. M_1 and M_2 are equivalent when they cannot be distinguished from each other by their output sequences. This statement is the basis of protocol conformance testing, where the equivalence of an implemented protocol and a protocol standard is examined (see Part III).

Example 4.3

The following state-transition table characterizes the finite-state machine M_1

STATE \ INPUT EVENT	0	1
1	2/0	6/0
2	3/0	5/1
3	4/1	2/0
4	5/0	1/0
5	6/0	4/1
6	2/1	5/1

Let us consider how the behaviour of M_1 changes on the effect of the input sequence 001, and which final state and which output sequence belong to which starting state and for which we can find equivalent states. In the state-transition table only those steps are indicated which have reactions on the input sequence. The step-numbers are marked with a circle (Table 4.3).

The last table contains the results. It is clear from the table of the results that every output sequence is different and the initial state by this input sequence is equal with the final state only at the initial state 5.

Table 4.3. Behaviour of M_1 depending on the input sequence 001 (Example 4.3)

	0	1
1	2/0 ①	
2	3/0 ②	
3		2/0 ③
4		
5		

OUTPUT: 000
FINAL STATE: 2
Table 1

	0	1
1		
2	3/0 ①	
3	4/1 ②	
4		1/0 ③
5		

OUTPUT: 010
FINAL STATE: 1
Table 2

	0	1
1		
2		
3	4/1 ①	
4	5/0 ②	
5		4/1 ③

OUTPUT: 101
FINAL STATE: 4
Table 3

	0	1
1		
2		
3		
4	5/0 ①	
5	6/0 ②	
6		5/1 ③

OUTPUT: 001
FINAL STATE: 5
Table 4

	0	1
1		
2		5/1 ③
3		
4		
5	6/0 ①	
6	2/1 ②	

OUTPUT: 011
FINAL STATE: 5
Table 5

	0	1
1		
2	3/0 ②	
3		2/0 ③
4		
5		
6	2/1 ①	

OUTPUT: 100
FINAL STATE: 2
Table 6

STATE	OUTPUT	FINAL STATE
1	000	2
2	010	1
3	101	4
4	001	5
5	011	5
6	100	2

State Identification with Successor Tree

If an input sequence occurs at the input of a finite-state machine, then an output sequence arises at its output and the state of the finite-state machine changes following each new element of the input sequence. We cannot observe the state transitions directly, but the output sequence is observable. Therefore we should work out a method to observe the finite state machine behaviour that observes the members of the output sequence as a reaction of the input sequence. The output sequence can characterize the behaviour of the finite-state machines unambiguously when we choose such input sequences that determine the states too. The 3-element-input-sequence in Example 4.3 produces a different output sequence in each state of the machine and from the result we can draw conclusions for every signal state.

The successor* tree is the basis of systematic state-identification. What kind of uncertainty exists when finite-state machines are examined?

If the finite-state machine M has n states, then initially it can be in any state. The initial uncertainty regarding the states of the machine can be described by the next state set, which is the set of possible next states

$$S = (S_1 \; S_2 \; \ldots \; S_n)$$

The initial uncertainty is the subset of S containing the possible initial states. If our machine M can be initially in any of four states, then the initial uncertainty is

$$(\,A\;B\;C\;D\,)$$

First of all by examining finite-state machines we want to reduce the initial uncertainty and to determine — if it is possible — the initial and or final states. The input sequence should be chosen either to eliminate the initial uncertainty or at least to reduce it. We can deduce from the response to the input sequence what the previous state was. Every different input sequence element means a branch of the tree. The input sequence is composed of so many members what every terminating state should be separated and the precedents should be inferred from the received output sequence. The number of elements of the input sequence is equal to the number of levels of the successor tree.

If the machine M_A has an input alphabet, x_1, x_2, x_3, x_4, then 4 branches start out from every node of the successor tree (Fig. 4.4).

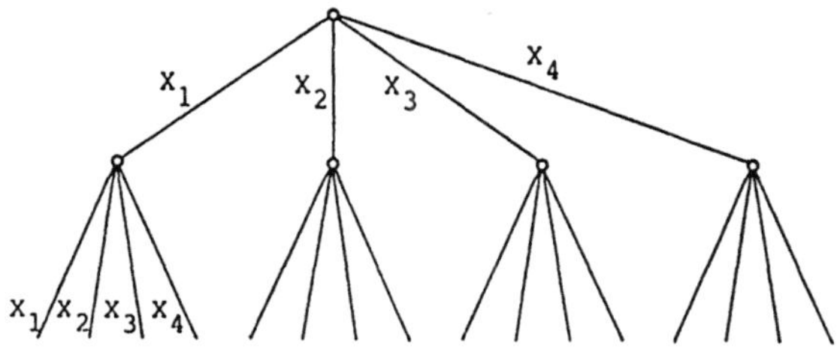

Fig. 4.4. Successor tree detail of machine M_A

* The word "successor" has a colloquial meaning: legal successor, heir, the next or the following. Here it refers to the states following each other during the state transitions.

The number of levels of the tree cannot, generally, be determined; it depends on the number of states and on the two characteristic functions: f_z and f_s or — more demonstratively — it depends on the state-transition table of the machine. To any specified machine M and to a given initial uncertainty belongs a successor tree demonstrating the successor uncertainty I_i for the whole input sequence I_i and helping to select the most suitable input sequence. The tree has branches whose levels following each other are denoted as 0, 1, ..., j. Every branch at the j-th level splits to branches. If the number of new branches is p, then these are labelled I_1, I_2, ..., I_p according to the input symbols. The branches starting from the j-th level create the j+1-th level. One uncertainty vector belongs to each node of the successor tree. The highest node (zero level) is characterized by the initial uncertainty U_o and every p node of the first level is coupled to the starting point with the successor of U_o. The j-th level of the tree is composed of the branches P_j and every branch terminates in a single node. The path of the tree is the series of branches starting from the highest node and terminating in one node of the j-th level. The length of the path is j. Every path describes an input sequence which, connected on the input, results in an uncertainty vector terminating on each node of the j-th level [KOH-78].

Example 4.4

Let us create the successor tree of the FSM M_1. The state-transition table of M_1 is Table A. Let us write the terminating states belonging to each state and the output states by the input sequences 00, 01, 11 and 111 (Table 4.4). Terminating states belonging to the same output sequence cannot be distinguished.

The four tables demonstrate well how the branches of the successor tree should be defined.

The states belonging to the same output are in brackets, these are not distinguishable. The input alphabet has only two elements and therefore the successor tree is equivalent with a binary tree (Fig. 4.5).

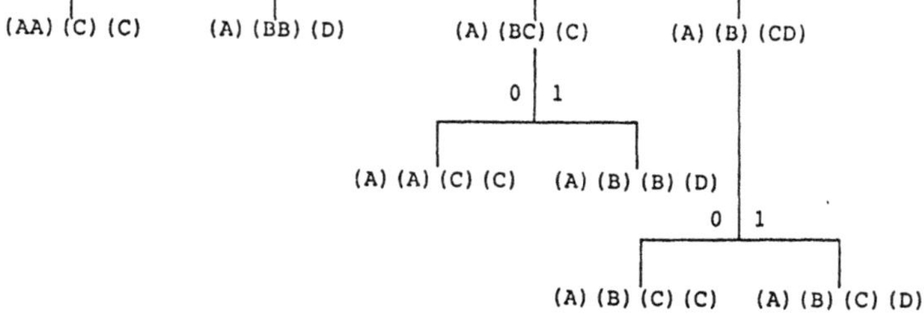

Fig. 4.5. Successor tree (Example 4.4)

Table 4.4. Construction of the successor tree (Example 4.4)

STATE-TRANSITION TABLE OF M_1

INPUT EVENT / STATE	0	1
A	C/0	D/1
B	C/0	A/1
C	A/1	B/9
D	B/0	C/1

STATE	OUTPUT	FINAL STATE
A	01	A
B	01	A
C	10	C
D	00	C

INPUT: 00

FINAL STATES: AA C C

Table 1

STATE	OUTPUT	FINAL STATE
A	00	B
B	00	B
C	11	D
D	01	A

INPUT: 01

FINAL STATES: A BB D

Table 2

STATE	OUTPUT	FINAL STATE
A	11	C
B	11	D
C	01	A
D	10	B

INPUT: 11

FINAL STATES: A B CD

Table 3

STATE	OUTPUT	FINAL STATE
A	110	B
B	111	C
C	011	D
D	101	A

INPUT: 111

FINAL STATES: A B C D

Table 4

It is clear from the successor tree that the branch according to Table 4 distinguishes each state, the sequence belonging to this path is called distinguishing sequence.

Example 4.5

Let us consider a simple application protocol. The input alphabet is

$$X = \{LOGIN, DATA, ACK, LOGOUT\}$$

and the output alphabet is

$$Y = \{DATA, ACK, LOGOUT, ERROR\}$$

and the set of states is

S = {OFF, WAIT, DATA, TRANSFER}

The state-transition graph and table in Fig. 4.6 describe the operation of the protocol (Table 4.5).

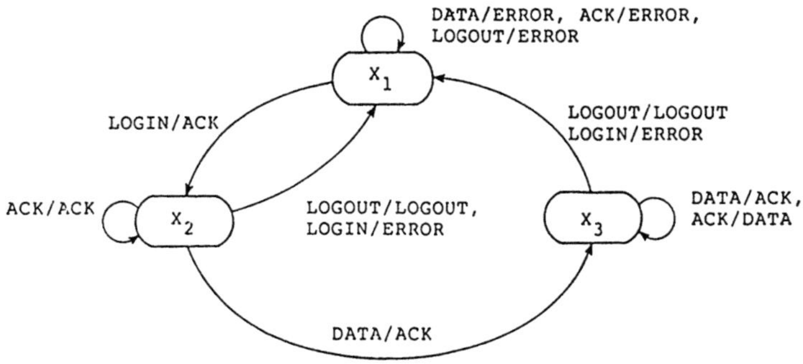

Fig. 4.6. State-transition graph of a simple protocol (Example 4.5)

Table 4.5. State-transition table for Fig. 4.6

State \ Input	LOGIN	DATA	ACK	LOGOUT
OFF X1	X2 / ACK	X1 / ERROR	X1 / ERROR	X1 / ERROR
WAIT X2	X1 / ERROR	X3 / ACK	X2 / ACK	X1 / LOGOUT
DATA TRANSFER X3	X1 / ERROR	X3 / ACK	X3 / DATA	X1 / LOGOUT

Let us determine the successor tree (Fig. 4.7).

$(X_1\ X_2\ X_3)$

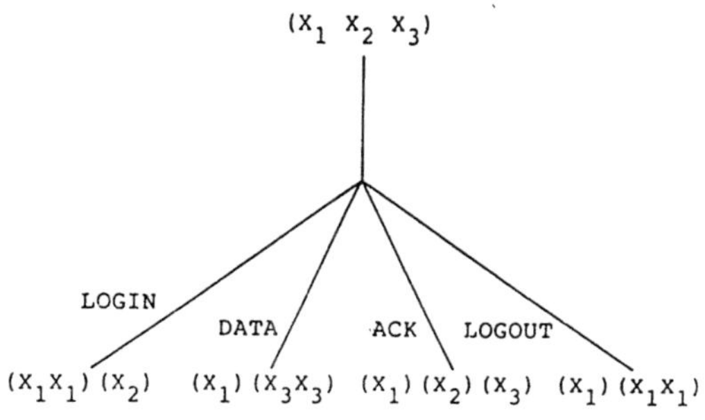

Fig. 4.7. Successor tree of a simple protocol (Example 4.5)

It is clear from the tree that it has no sense to continue the paths because we achieved the distinguishing sequence on the ACK branch in the very first step:

$$D = ACK$$

[BJOR-70], [BOCH-76].

5. GRAPH MODELS

5.1 PETRI NETS

Fundamental Concepts of Petri Nets

Petri nets were developed for illustrative description of the operation of complex systems. Carl Adam Petri suggested the basic model in his dissertation "Kommunikation mit Automaten" in 1962. The model [PET-77], [PET-81], [PETRI-62] has the following parameters:

— events,

— conditions waiting for fulfilment,

— fulfilled conditions.

Later Anatol Holt modified the basic model. This new model to be described below is easier to apply and has a broader range of application than the original one.

The relationship between the events and conditions of the system is described with two kinds of graphic elements:

— place

notation: 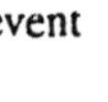

meaning: condition

— transition

notation: 

meaning: event

The places and transitions are connected with directed arcs. The places and transitions follow each other in alternating sequence [AGE-79], [DAN-77], [DAN-80], [POS-74].

The *place* can be empty or can have a token. The empty place represents a condition waiting for fulfilment, the place with a token is a fulfilled condition. The place with a token is known as a marked place, too. The tokens can migrate and wander in the graph. The ***marking*** of the graph means the distribution, placement of the tokens. The initial marking is the initial distribution of the tokens.

Petri introduces one token for one *condition*, Holt permits any number of tokens. The tokens can move, they can go from one place to another through the intermediate transition. Crossing a transition is named *firing*, actualization of the transition. A Petri net can have only a single firing at one time.

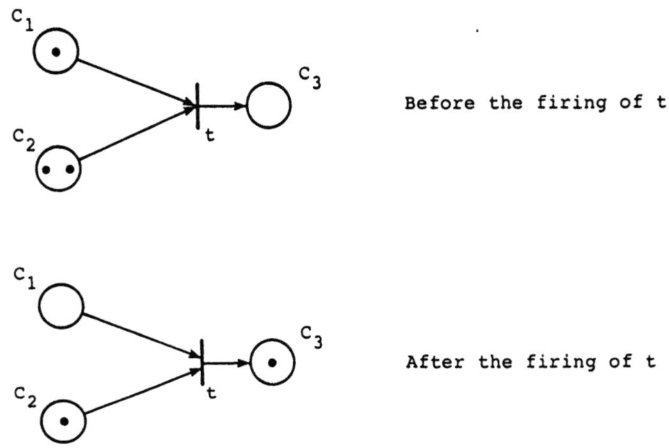

Fig. 5.1. Petri net

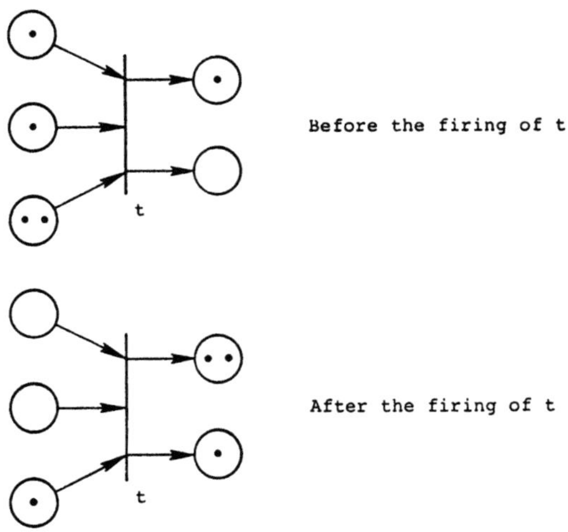

Fig. 5.2. A Petri net example

Firing rules

— a transition is in fireable state if every input place belonging to this transition has at least one token

— during the firing one token leaves each input place and one token enters into each output place.

The following examples demonstrate the firing rules (Figs. 5.1 and 5.2).

Petri Net Features

A firm marking can be

— reachable,

— bounded,

— safe.

A marking M_m is said to be *reachable* from a marking M_o if there is a σ firing sequence making it possible to reach M_m from M_o in a finite number of steps. The reachability set R(M) of the marked Petri net is the set of all markings reachable from M.

The *bounded* marking is an important characteristic of the graphs composed of places with more tokens. The initial marking M_o is bounded if any place reachable from M_o has only a k-number token.

The marking M_o is *safe* if $k = 1$. A marked Petri net is k-bounded if its each place is k-bounded and it is safe if each of its places is safe.

The t transition is in the M-marked Petri net when in each $M' \in R(M)$ there exists a marking reachable from M' activating the t transition (Fig. 5.3)

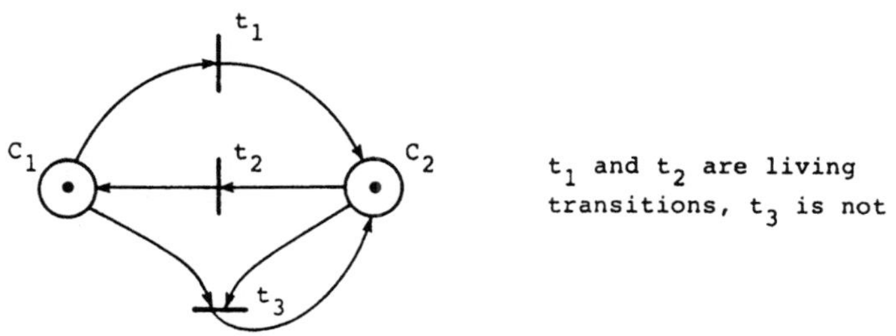

Fig. 5.3. Petri net

A Petri net can contain deadlocks and traps. These are nearly opposite concepts. The *deadlock* is a sort of the places surrounded with transitions and a token can enter into this part only if the deadlock subgraph contributes at least one token.

If a deadlock is unmarked, i.e. has no tokens, then it remains unmarked for each firing sequence.

The *trap* is such part of places of the Petri nets surrounded with transitions and if a token moves off then at least one token returns to the trap-subgraph. If a trap was marked, i.e. the trap-graph had at least one token, then it remains marked for every firing sequence.

The trap and deadlock do not preclude each other. If a deadlock has a marked trap, then it never becomes empty and it always remains living (Figs. 5.4 and 5.5).

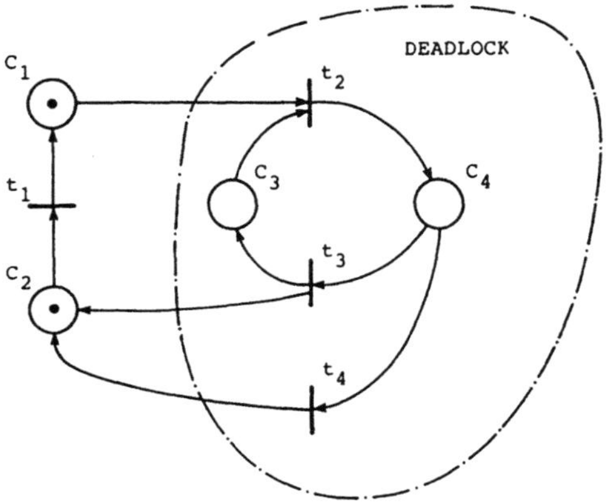

MARKING	c_1	c_2	c_3	c_4	t_1	t_2	t_3	t_4
M_0	•	•	•	0				
						GY		
M_1	0	•	0	•				
					GY			
M_2	•	0	0	•				
							GY	
M_3	•	0	•	0				
						GY		
M_4	0	•	0	•				

Fig. 5.4. Deadlock

Notation:

• *means token*

0 means empty place

GY is firing

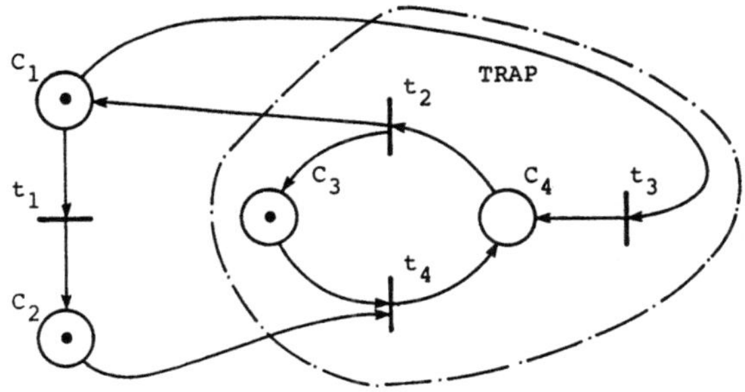

MARKING	c_1	c_2	c_3	c_4	t_1	t_2	t_3	t_4
M_0	•	•	•	0				
								GY
M_1	•	0	0	•				
					GY			
M_2	0	•	0	•				
						GY		
M_3	•	•	•	0				
								GY

Fig. 5.5. Trap

Notation:

- • *means token*
- *0* *means empty place*
- *GY* *is firing*

The Subgraphs of the Petri Net

A Petri net can be decomposed to different subgraphs. The consistent subgraphs have two classes:

— closed,

— open.

The *closed consistent* subgraph contains a set of places and every transition coupled to these places is independent from the arc directions. The deadlock and trap belong to this class.

The *open consistent* subgraph contains a set of transitions and every place is coupled to these transitions independently of the arc directions.

The most frequently used subgraphs are the following (Fig. 5.6).

a) STATE MACHINE graph
   Every transition has a single input place and a single output place
b) MARKER graph
   Every place has a single input transition and a single output transition
c) FREE CHOICE graph
   Every arc starting from a place is either a single output of a place or a single input of a transition
d) SIMPLE PETRI nets
   Every transition has a maximum of one shared place.

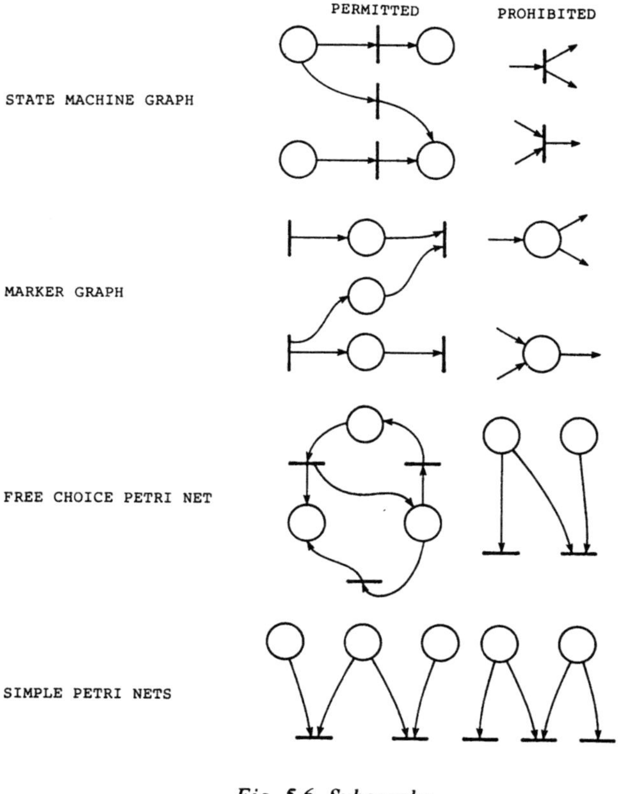

Fig. 5.6. Subgraphs

Mathematical Description of Petri Nets

A Petri net is a four-tuple and a directed graph

$$N = (T,P,A,M_0)$$

where

T is the set of transitions (events)
P is the set of places (conditions)
A is the set of directed arcs
M_0 is the initial marking

None of the above sets is empty. The physical interpretation of the transitions is events. A set of transitions is given by

$$T = \{t_1,t_2, \ldots t_n\}$$

The physical meaning of the places is conditions. A set of conditions is given by

$$P = \{p_1,p_2 \ldots p_m\}$$

The graph points are

$$T \cup P$$

A set of directed arcs is the relation assigning the set of places to the directed arcs of the graph.

The set of input places belonging to transition t is

$$I(t) = \{p \mid (p, t) \in A\}$$

namely those input places belong to the transition t whose arcs (p, t) are the elements of set A.

The set of output places belonging to transition t whose arcs (t,p) are the elements of the set A is given by $O(t) = \{p \mid (t,p) \in A\}$.

The interpretation of the input and output places is illustrated in Fig. 5.7.

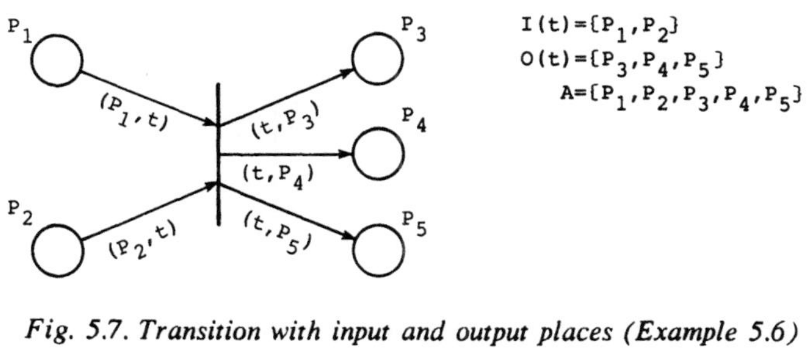

Fig. 5.7. Transition with input and output places (Example 5.6)

A transition t with marking M can fire when $M(p) > 0$ at each $p \in I(t)$.
The incidence matrix describes how the arcs are fitted to the places [AND-83].
Incidence matrix:

$$C = \begin{array}{c} \\ t_1 \\ t_2 \\ \vdots \\ t_n \end{array} \begin{array}{c} \begin{array}{cccc} p_1 & p_2 & & p_m \end{array} \\ \left[\begin{array}{cccc} c_{11}(t,p) & c_{12}(t,p) & \cdots & \\ \cdots\cdots & c_{22}(t,p) & \cdots & \\ \cdots & \cdots & c_{jk}(t,p) & \cdots \\ \cdots & \cdots & \cdots & c_{nm}(t,p) \end{array} \right] \end{array}$$

The general element of the matrix:

$$c_{jk}(t,p) = \begin{cases} -1 & \text{if } (p,t) \in A \text{ and } (t,p) \notin A \\ 1 & \text{if } (t,p) \in A \text{ and } (p,t) \notin A \\ 0 & \text{else .} \end{cases}$$

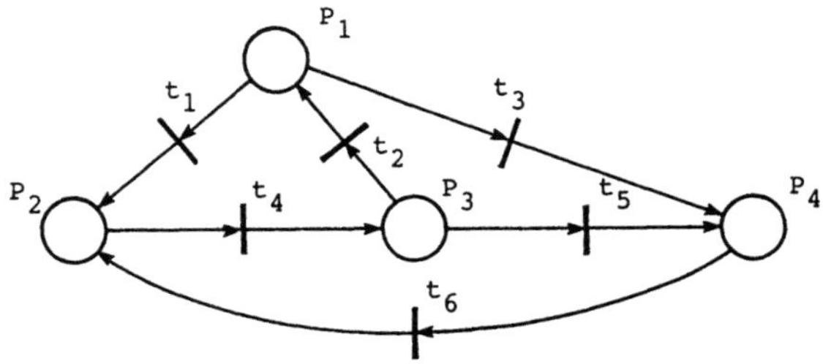

$$C = \begin{array}{c} t_1 \\ t_2 \\ t_3 \\ t_4 \\ t_5 \\ t_6 \end{array} \overset{\begin{array}{cccc} p_1 & p_2 & p_3 & p_4 \end{array}}{\left[\begin{array}{cccc} -1 & 1 & 0 & 0 \\ 1 & 0 & -1 & 0 \\ -1 & 0 & 0 & 1 \\ 0 & -1 & 1 & 0 \\ 0 & 0 & -1 & 1 \\ 0 & 1 & 0 & -1 \end{array} \right]}$$

Fig. 5.8. Petri net with its incidence matrix (Example 5.7)

Figure 5.8 demonstrates the relationship between the Petri net and the incidence matrix.

Finally the way of the tokens can be seen (Fig. 5.9) from the initial marking again into the starting marking [BER-TER-81], [DIAZ-82].

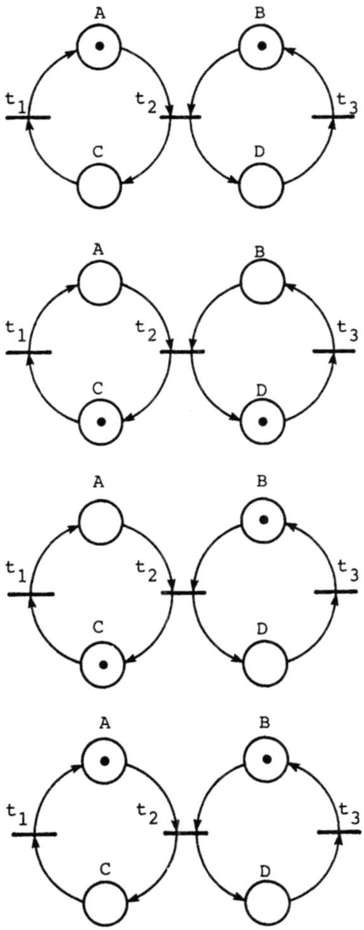

Fig. 5.9. Petri net (Example 5.8)

5.2 MODIFIED PETRI NETS

Classification of Modified Petri Nets

Petri nets are demonstrative, graphic descriptions of the systems serving to couple events with conditions. The elements of the basic Petri net are shown in Fig. 5.10.

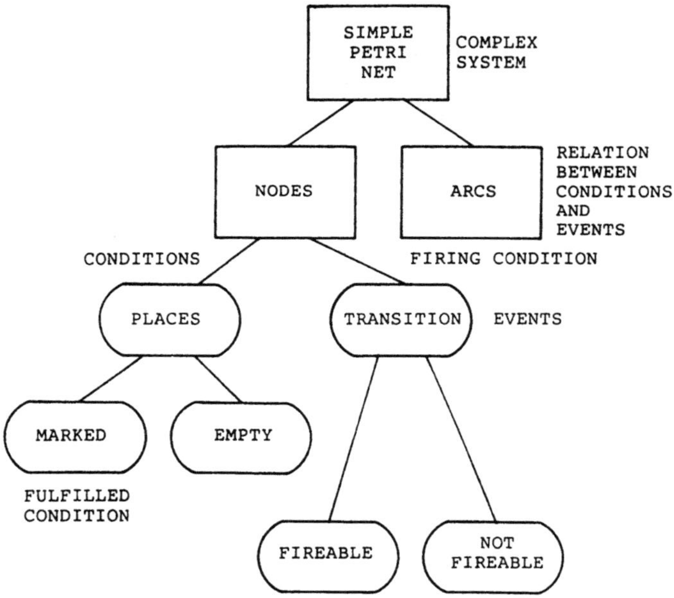

Fig. 5.10. Elements of the basic Petri net

Modified versions of Petri nets originated in order to solve some difficulties arising during the course of system modelling. The main problem is to interpret graphically the synchronization of events and the sequencing of the decisions. Therefore the properties of the places, transitions and arcs should be changed, the tokens should be augmented with new parameters to represent these problems clearly and unambiguously [NAGY-76].

The following modifications are most frequently used:

— time Petri net,
— E-graph,
— numerical Petri net.

The *time Petri net* (Fig. 5.11) makes the firing of the transitions more deterministic, it assigns a time interval within which a transition must fire. In this way the synchronization of event sequences is possible.

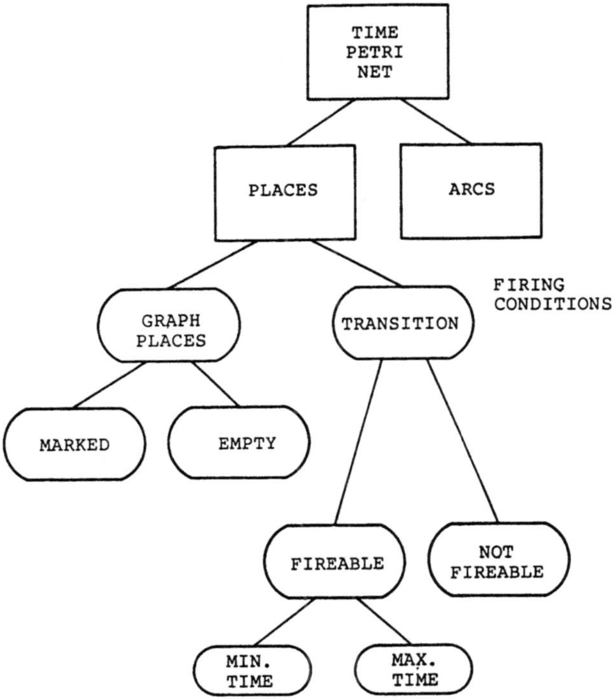

Fig. 5.11. Elements of time Petri net

The *E-graph* (Fig. 5.12) simplifies the determination of the directions of logical decisions introducing decision places and ensures the synchronization with time values assigned to the transitions.

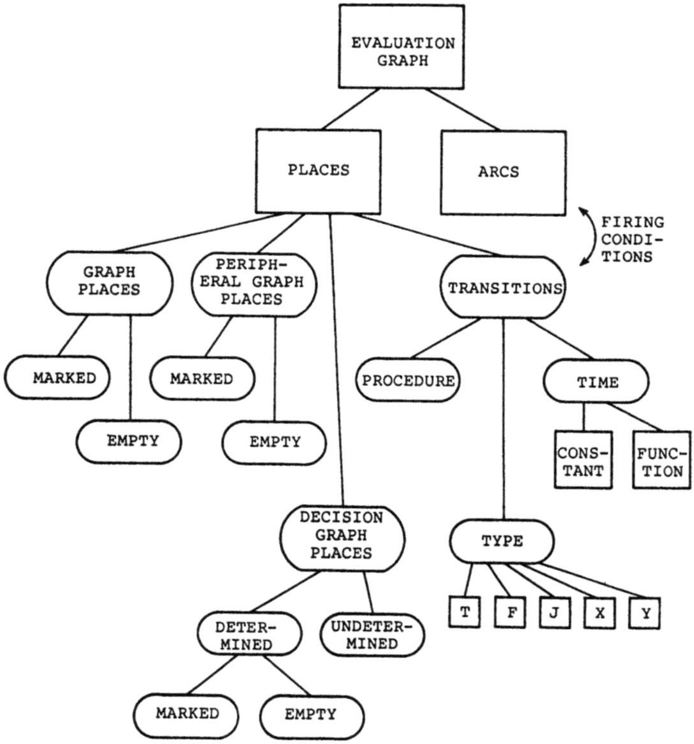

Fig. 5.12. Elements of the E-graph

The *numerical Petri nets* (Fig. 5.13) complete the relations between the transitions and arcs with preparation and firing conditions and allocate memory and operation rules to the transitions.

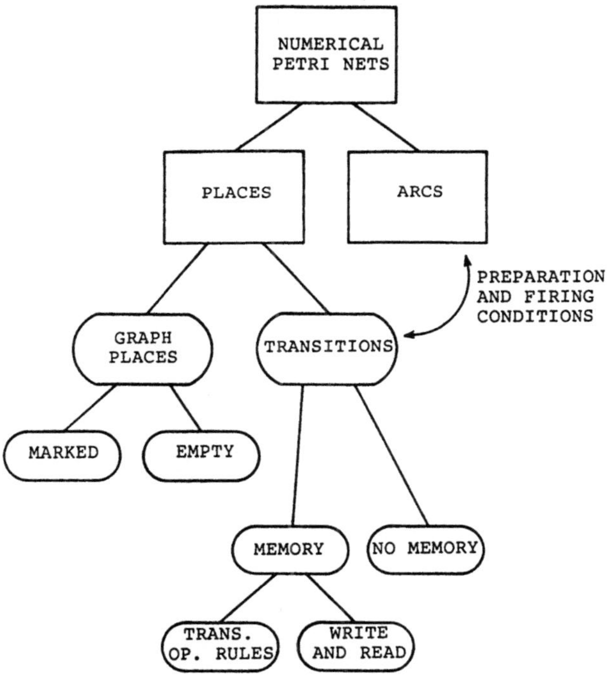

Fig. 5.13. Elements of numerical Petri net

Time Petri Nets

The time Petri net is the simple extension of the basic Petri net. It is important in numerous tasks that a condition is fulfilled within a defined time-interval.

Philip Merlin suggested the completion of the basic Petri net with time parameters in 1976 [MER-76]. This modification makes it easier to build timing and synchronization into the model.

Merlin made a proposition to assign the values t_{min} and t_{max} to every transition. If each input place has one token, then the time period must be after the appearance of the last token and before the firing more than t_{min} but less than t_{max}.

The basic Petri net is a special case of the time Petri net. Every transition has the same time parameters

$$t_{min} = 0$$
$$t_{max} \rightarrow \infty$$

Figure 5.14 demonstrates the operation of the time Petri net well. When the event t_1 occurs within the $\Delta t_1 = t_{1\,max} - t_{1\,min}$ interval, the transition t_1 fires and a token enters into the place A. After that, t_2 fires and B becomes marked. The token remains in B until C becomes marked, t_4 fires with Δt_3 after the marking of C and the token goes to D. At the right side of the figure the (in place) elapsed time of the token can be seen.

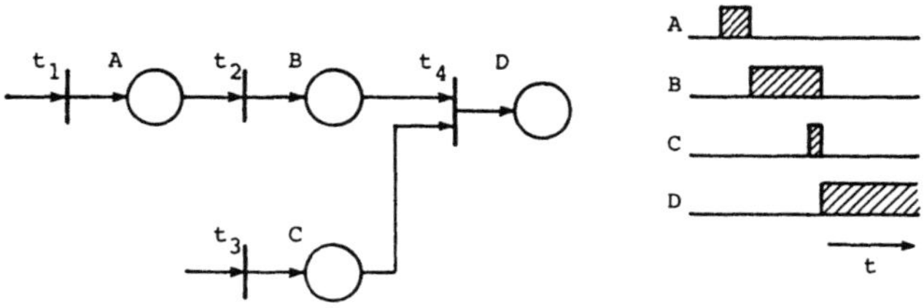

Fig. 5.14. Operation of the time Petri net

It may be noted that the above introduced Merlin time parameters differ from the time concept used in E-graphs.

Some examples for time Petri nets are given in Section 5.4.

E-nets

Evaluation nets (E-nets) were first proposed in 1972 by Nutt. By introducing E-nets he was able to eliminate the indeterminism of Petri nets. He introduced for the exact description of decisions connected with concurrent events the notion of decision net-place and connected a time parameter to every event. (Unfortunately the Merlin and the Nutt time parameters are different.) [NUTT-72]. Evaluation nets can be defined by ordered 5-tuples:

$$E = (T_A, P, P_p, R, M_o)$$

where:

- T_A is the set of transitions
- P is the set of net places
- P_p is the set of peripheral net places and $P_p \subset P$
- R is the set of decision net places and $R \subset P$
- M_o is the initial marking of a net place

No sets can be empty except the set of peripheral net places.
E-nets could have three different types of net places:

— Simple net places 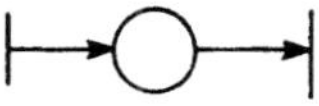 which are identical with the Petri net places.
— peripheral net places
  • if the transition has only input place(s) 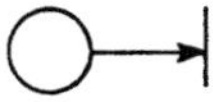
  • if the transition has only output place 
  • decision net places  which determine the direction of token movements.

The state of a simple net place and that of peripheral net places are always definite: Marked (it contains a token), or empty (it does not contain a token). But decision net places can also have a third and undefined state. The states of the decision net places are described by r control functions. The value of r is equal to 0, 1 and the value of r determines whether a token enters or leaves a given net place. If the value of r is undefined, then no token movement is allowed. One net place must not contain more than one token, this means that an E-net is always safe.

Markings of decision net places

$$M(r) = \begin{cases} 1 & \text{if the predicate p is true} \\ \emptyset & \text{if the predicate p is false} \end{cases}$$

Predicate* selection will be discussed later in the application chapter.

The definition of transitions in E-nets supports also efficiently the deterministic description of the system. The i-th transition can be described by an ordered three-tuple:

$$a_i = (s,t,q)$$

where

s is the type of transition
t is the time required for a transition
q is the transition process

The type of transition can be (Fig. 5.15)
— T-transition,
— F-transition,
— J-transition,
— X-transition (it also has a decision place),
— Y-transition (it also has a decision place).

* Predicates express the characteristics and qualities of things or connections and relationships between two or more things.

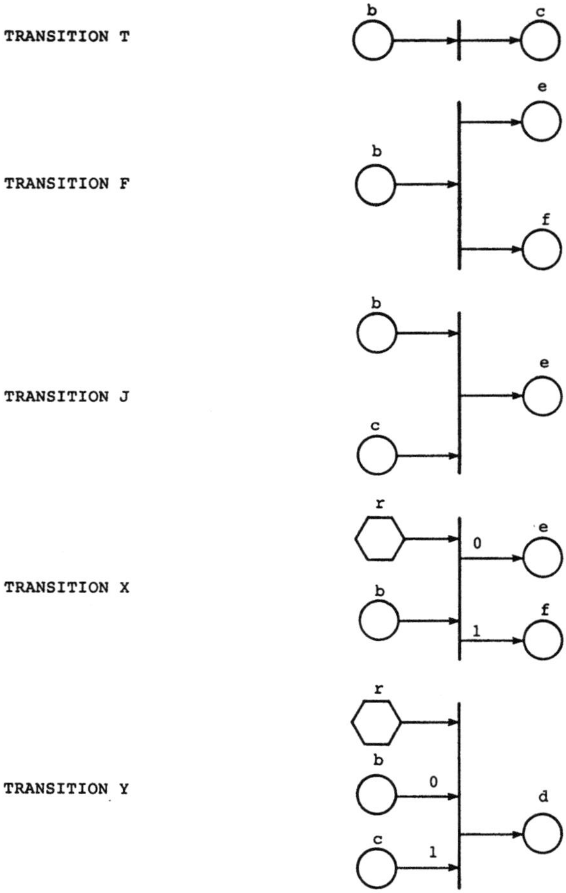

Fig. 5.15. Type of transitions

T-transitions have one input and one output net places. If a token enters net place b and net place c is at the same time empty, then the transition fires in time f and the token moves from b into c.

F-transitions have one input and two outputs. If a token enters net place b and net places e and f are empty, then the transition fires and the token moves into net places e and f.

J-transitions have two inputs and one output. J-transitions fire, if b and c are marked and e is empty. Tokens leave both input net places and enter the empty net place e.

X-transitions are similar to T-transitions, but contain decision net places too. The output edges of an X-transition are marked by 0 and 1. If the state of r is indefinite nothing happens. If r is equal to 1 and a token enters the net place b, then the transition fires and the token moves into net place f. At the end of the firing b will be empty and the state of r becomes indefinite again.

Y-transitions are similar to J-transitions, but contain decision net places too. If the value of r is undefined, then the transition cannot fire, however tokens leave those net places which are marked by the decision function r [MAR-82].

By the ordered 3-tuple, it has been *transition time* correlated to the i-th transition. The total transition time is the sum of the initial waiting time and the firing time. The initial waiting time is the maximum time spent by one of the tokens at the input net place until it has fired. In the case of a transition having more than one input net place, that token's waiting time dominates which had first entered an input net place.

Firing periods are the following:

- Pseudo fireable period
  there are tokens in all the net places, but the state of the decision net place is undefined
- fireable period
  in all the net places there are tokens and the states of decision net places are also definite, so the transition will fire
- active period
  the transition can be fired and is already in action, but the states of all the related net places are so far undefined
- final period
  tokens are leaving the input net places and entering the output net places, so the transition process has been finished.

The system described has become completely *deterministic*, since times are assigned to every event (every transition) and the transition's firing sequence is unambiguously determined.

Transition times can be given as constants or functions. If transition times are constants, then they correspond to the time interval between the appearance of tokens at the input net places of the transitions and the firing of the transition. If a transition time has been defined as a function, then this function can be described by an attribute vector or global variable.

The attribute vectors are associated with the tokens and describe their characteristics. Attribute vectors might change when the tokens move.

Let us consider the case when a token has at place **a** an n-element and at place **b** an m-element attribute vector. If n is greater than m, then the last n-m elements are lost. If m is greater than n, then the last n-m elements are undefined.

It is common to apply the limitation that a token's attribute vectors in all the net places of an E-net must have the same numbers of elements (n=m).

The vector of the global (peripheral) variables are associated with the whole system. The function will have at every transition a different value. This function is given by the third component of the E-net's transitions, by the ***transition procedure***. The components of the transition procedure are usually arithmetic expressions.

Numerical Petri Nets

The numerical Petri net is another generalization and extension of the classical Petri net. The interpretation of the places and the fundamental operation principle (only a single transition can fire at one time) are unchanged, but the transitions and tokens have many more optional parameters such as: enabling conditions, memories, operation rules [SYM-80], [SYM-81], [WHE-SIM-82].

The differences are the following:

- the fulfilment of firing conditions is not enough for the firing of a transition: the enabling conditions should be fulfilled, too
- different firing conditions belong to the input and output arcs of the transitions
- the transitions may have memories
- different transition operation rules are assigned to the transitions
- the tokens are characterized either by a natural number or by an attribute-vector.

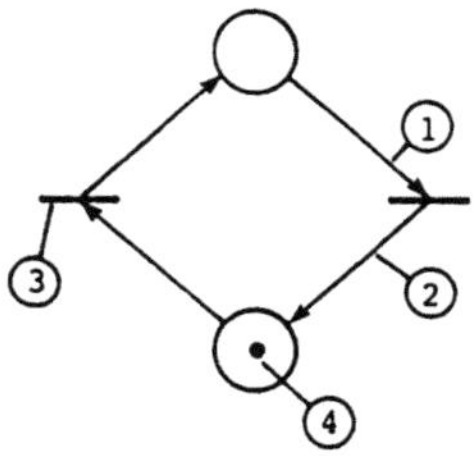

1 to the input arc of the transition belong:
- firing condition (in)
- preparation condition

2 to the output arc of the transition belongs:
- firing condition (out)

3 a) the transition has
- write memory
- read memory

b) the operation rules of the transition determine the operation on memory data to be executed

4 token is
- a natural number
- an attribute vector

Fig. 5.16. Example of numerical Petri net

A simple numerical Petri net can be seen in Fig. 5.16. The input arc of transition t_1 has an input firing condition and an enabling condition. The output arc of transition t_2 has an output firing condition. Transition t_2 has a memory and the operation rules determine the actions to be executed depending on the data in the memory. Token T has a natural number and an attribute-vector.

The numerical Petri net has some interesting firing rules. As already discussed, a transition can fire only when the enabling conditions as well as the firing conditions are fulfilled. Both conditions are Boolean. The fulfilment of the two conditions is independent. The firing condition belonging to the input arc is concerned with the tokens leaving the appropriate input places at the firing of the transition. The firing condition belonging to the output arc controls the token entering into the predetermined place after the firing.

There are reference data in the memory of the transition. The enabling conditions define the writing of the reference data. The new reference data enter into the memory by firing, the actions according to the transition operation rules are executed on the data, then the tokens leave the transition and the temporary memory is cleared.

The parameters and conditions assigned to the numerical Petri net are demonstrated in the following simple example (Fig. 5.17). In the example we use the following notation:

- The name of places A, B, C, D
- The tokens of the places G, H
- The transition names t_1,t_2,t_3
- Enabling conditions E_1,E_2,E_5,E_7
- Firing conditions
  - F_1, F_2, F_5, F_7
    belonging to input arcs
  - F_3, F_4, F_6, F_8
    belonging to output arcs
- The operation rule of the transition
  Y: = F (x)
- The memory reference condition
  [Z = K]

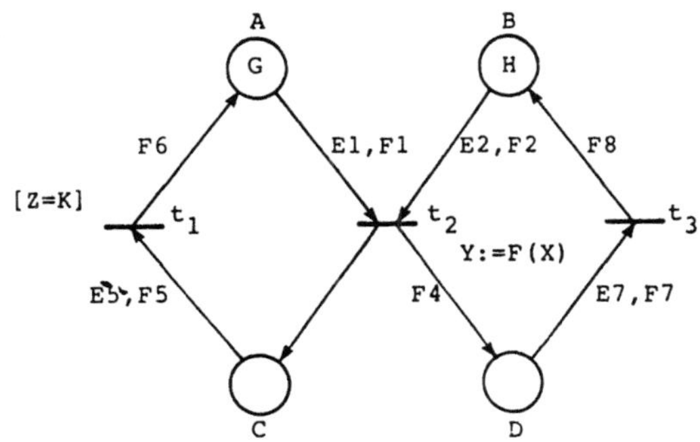

Fig. 5.17. Simple numerical Petri net

5.3 DATA FLOW GRAPHS

Graph Model of Behaviour

Gardner summarized the graph models of computer science in his dissertation in 1975 and suggested the application of the Graph Model of Behaviour (GMB). GMBs comprise three parts [RAZ-ES-82]:

— control flow graph,

— data flow graph,

— interpretation.

A *control flow graph* has nodes and arcs connecting the nodes. The nodes represent events, and the arcs are the relations between the events.

Input and output logical expressions (AND or OR) are assigned. The input operations concern the arcs running in to the node and determine that to activate this node every arc (AND operation) or one or more arcs (OR operation) should be in the controlled state. The output operations allocate either every outgoing arc (AND operation) or only a part of the arcs (OR operation) into the controlled state.

The control flow graph is the static description of all possible control sequences. The actual state of the graph is determined by the tokens assigned to the arcs. The activated node sinks the incoming tokens, but allocates new tokens to the outgoing arcs according to the output operations.

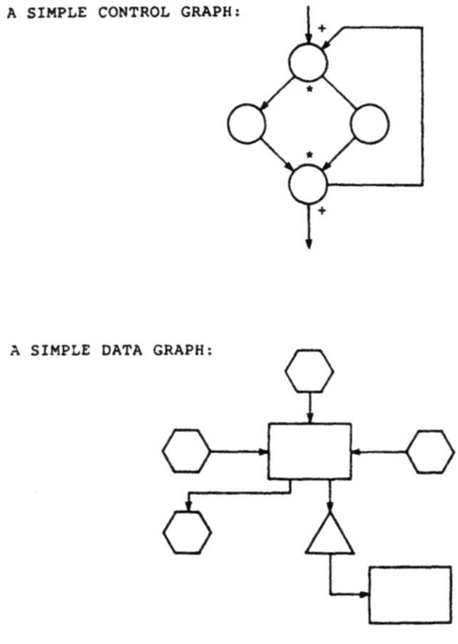

Fig. 5.18. Control and data graph

A simple example of a control flow graph is given in Fig. 5.18.

A *data flow graph* is composed of:

— data sets,

— controlled processors,

— uncontrolled processors,

— data arcs.

The data sets collect the data, and the processors execute data transformations. The controlled processors reach, by data transformation, the initialization of the control graph. The uncontrolled processors reply with data transformation on the changing of the input data.

The data arcs connect the data sets and processor. An example for graph model of behaviour can be seen in Fig. 5.19.

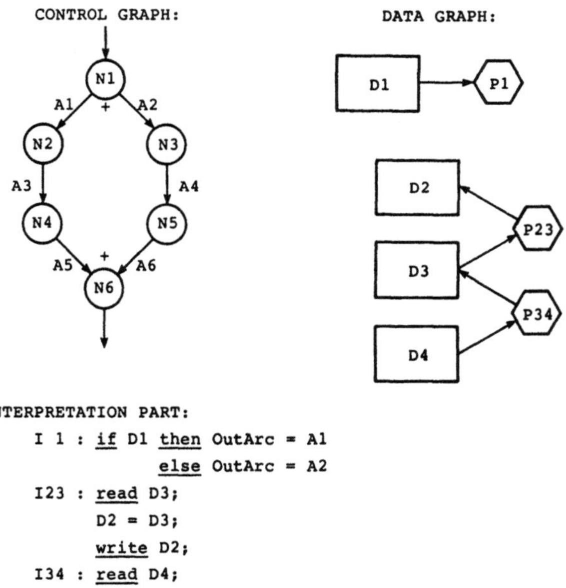

INTERPRETATION PART:

```
I 1 : if D1 then OutArc = A1
           else OutArc = A2
I23 : read D3;
      D2 = D3;
      write D2;
I34 : read D4;
      D3 = D4;
      write D3;
```

Fig. 5.19. Example of the graph model of behaviour

What kind of further knowledge is added by the *interpretation* to the system information described by the flow control graph and data flow graph?

— it determines the format of the data stored in the data sets,

— it prescribes the data transformations to be executed by the processors,

— it enumerates the decisions to be made by the control graph.

Generally high-level languages are used for the interpretation.

There is an unambiguous correspondence between the state transition graphs and the graph model of behaviour. The advantage of the GMB is that it preserves the illustrative power of the graph methods but it prevents the state explosion and the size of the graphic representation because of the decomposition of the system description into three connecting parts [AHO-SAR-87a], [BU-SA-69].

Data Flow Graph Models

In traditional serial execution the order of the instruction execution is the same as the order of the instructions in the program. Opposed to this is the data flow model in which the programmer is not dealing explicitly with the order of the execution. His only intention is to preserve the correctness of data relationships because the partial results are the initial data of the further computing.

The computation to be performed is specified by means of the *data flow graph* composed of the elementary computing process and their relations.

As an example, the determination of the product of two complex numbers is demonstrated in Fig. 5.20.

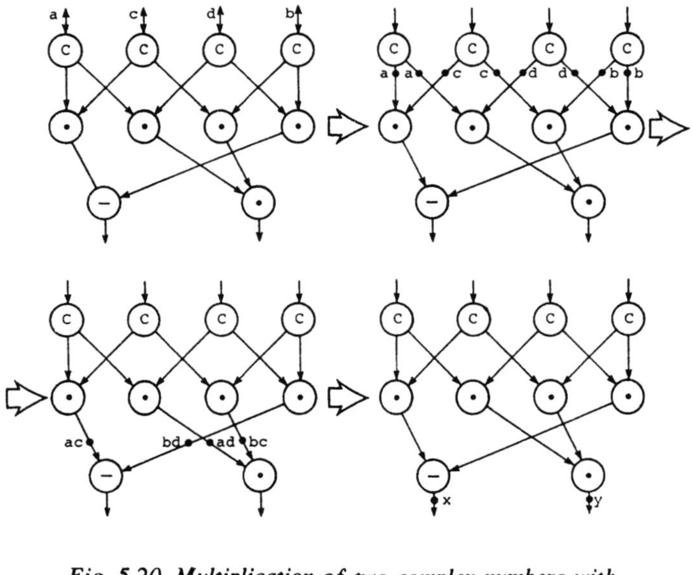

Fig. 5.20. Multiplication of two complex numbers with data flow graph

The data flow graph represents a two-dimensional structured program.

Every instruction execution is assigned to a place; the arcs represent determined data relations.

The data values as tokens or points cross the directed arcs of the graph during the program execution, while the execution of operations in the vertexes, nodes are executed. The tokens arrive to the node, then it is activated, ready to fire. The process uses up the tokens and carries out the appropriate operation resulting in new data tokens on the output arcs. The process is similar in every place.

How can this task be decomposed into well defined steps?

The starting point is the fact that any computation can be decomposed into elementary operations (these are the operators), and these are connected to each other through datapaths. The datapaths deliver the data values and at the same time store them.

The data flow program can be considered as a directed graph, in which the places specify (asynchronous elementary executed) operators.

The directed arcs regard the dataways determining the possible data flow.

The operators have input-output arcs and define such elementary operations as addition, multiplication, conditional data transferring, etc. on the input values; the result is always replaced on the output arcs.

There are many data flow concepts and models, but three principles are common for all of them.

- The data flow operation is executed if and only if every operand of the operation is computed.
- The data values disappear from the operator input after the execution of the operation.
- The computing is based on the computed values and not on the positions where they can be found: the operations cannot preserve internal state-information.

The basic models realize the basic principle, but they differ from each other in the type of the defined operators, in their complexity, in the handling of the data structure and in solving the re-entrant problem.

Dennis and his colleagues [DEN-80] carried out pioneering work in the field of the data flow computing, the operators have four defined basic types in their model:

- PCF computing function,
- SWITCH,
- MERGE gate,
- APPLY.

The APPLY operator is an abstraction of the procedures of the traditional programming languages: it handles the user-defined data flow structure as an operator. The firing rules of the operators can be seen in Fig. 5.21.

Behcet Sarikaya successfully utilized a data flow graph for protocol specification and testing [SAR-84], [DEN-80].

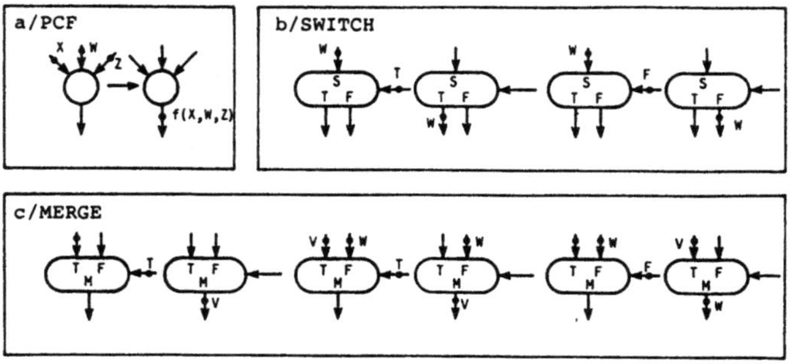

Fig. 5.21. Firing rules of operators

5.4 EXAMPLES OF GRAPH MODELS

Examples of Petri Nets (Fig. 5.22)

E x a m p l e 5.1 shows simple logical branching. If an input event occurs, then transition T1 fires and a token is moved into place G1. After this, either event S1 or S2 occurs randomly because only one transition can fire in a Petri net at one time. Therefore the token goes from G1 to G2 either through transition S1 or S2; that is, condition G2 is fulfilled. Finally transition T2 fires and the output event T2 is followed.

In **E x a m p l e 5.2** the modelling of a competition is demonstrated.

Let us have the following initial marking:

p_1 is marked,

p_2, p_3, p_4 and p_5 are empty.

After the firing of t_1, p_2 and p_3 become marked p_1 although empty.

Next, either t_2 fires first and t_3 the second time or reversely. If both transitions were fired, then both input place of t_5 are marked and t_5 can fire. The token returns to p_1. It is one of the possible parallel processes. The other possibility is that p_4 has a token and t_4 fires, the token returns to p_2. The token circulates in this loop until the firing of t_5 precedes the firing of t_4.

E x a m p l e 5.3 demonstrates a more complex logical problem. The initial token is in p. After firing either S1 or S2, t_1 will fire and the firing of transitions S_3, S_4 and S_5 in the three parallel branches take place randomly. If the transition t_3 has tokens in all three input places, then t_3 fires and the token moves into place K. If t_1 fires earlier, then the token goes to L and from here back to K through the firing S_0. If t_2 fires earlier, then the token arrives to p and in this way the marking is the same as initially.

In **E x a m p l e 5.4** a processor has two I/O devices. There is always a token in the place p_5 representing the accessibility of the processor. Event t_1 means the new data on the first device input, and event t_3 means the new data on the second device input. The fulfilment of conditions p_1 and p_3 signifies the end of the transmissions of the previous data. The fulfilment of p_2 and p_4 signifies that the processor is ready to receive new data.

The model demonstrates well that the processor processes the single data at once, but if data are waiting in both branches then it picks up one of the two randomly.

How can we model that the processor should deal with the two inputs alternatively? There are various methods for this, one simple possibility being given in Fig. 5.22. The token alternates between p_5 and p_6 and therefore both inputs can send data alternately. But it is necessary to have tokens in both branches otherwise the enabling condition is never fulfilled.

E x a m p l e 5.5. The next Time Petri net realizes a threefold cycle. The firing time of the transitions is larger in the required order.

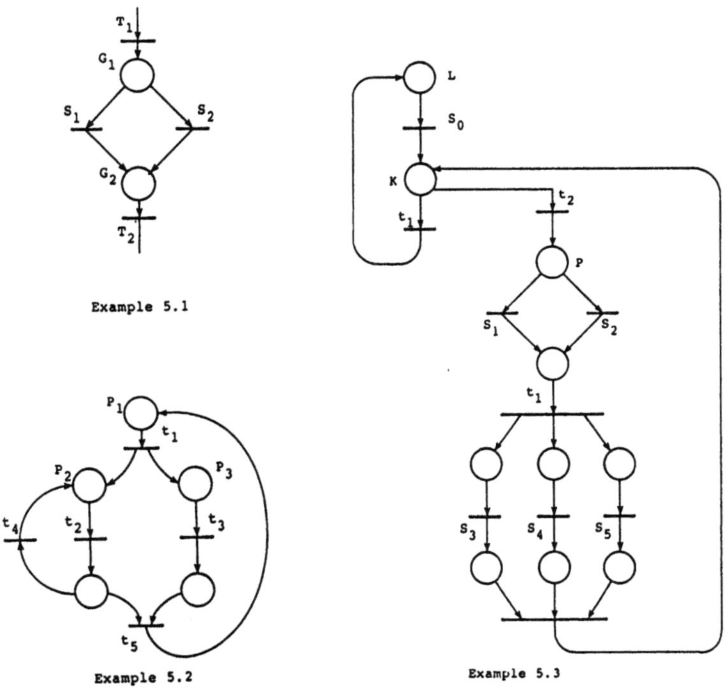

Fig. 5.22. Examples of Petri nets

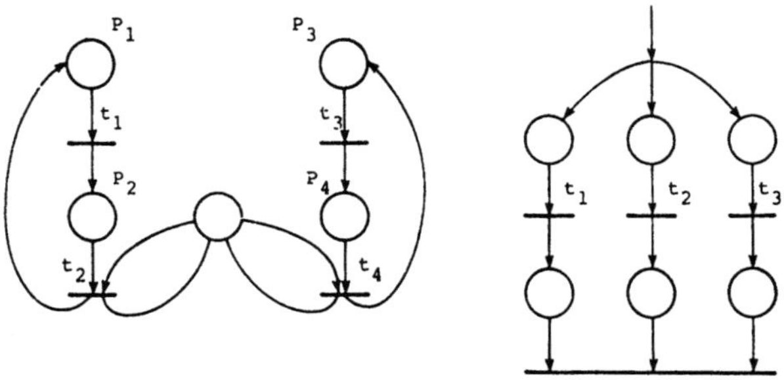

Example 5.4 Example 5.5

Fig. 5.22. Continued

Graph Models of Protocol Components

In **E x a m p l e 5.6** a sender and a receiver are connected to each other through a transmission medium with loss. The sender can send messages when the receiver has already accepted the previous message and the transmission medium is empty. If the token is in the place "MESSAGE ON THE WAY", then the token can either go from here to the receiver which is "RECEIVER READY", or it will be sunk by the place "MESSAGE LOSS" and the transmission medium will again be empty. The sender receives no acknowledgements from the receiver in this model (Fig. 5.23).

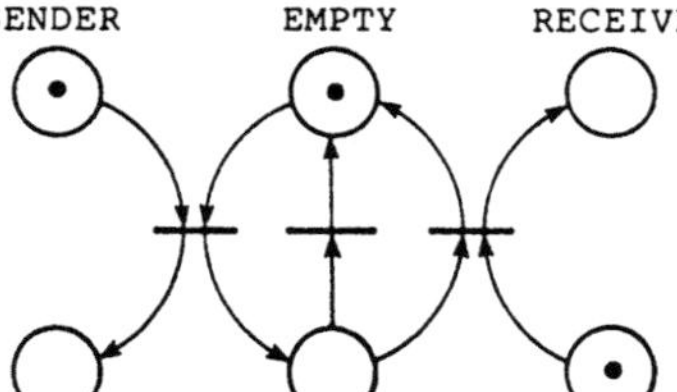

Fig. 5.23. Transmission medium with loss

E x a m p l e 5.7. The Time Petri net can demonstrate well the recovery mechanism following a lost message. The token disappearing from place M represents the transmission medium with loss. Transition t_7 inserted between M and W signifies the recovery mechanism. The notation is the same as in the first example (Fig. 5.24).

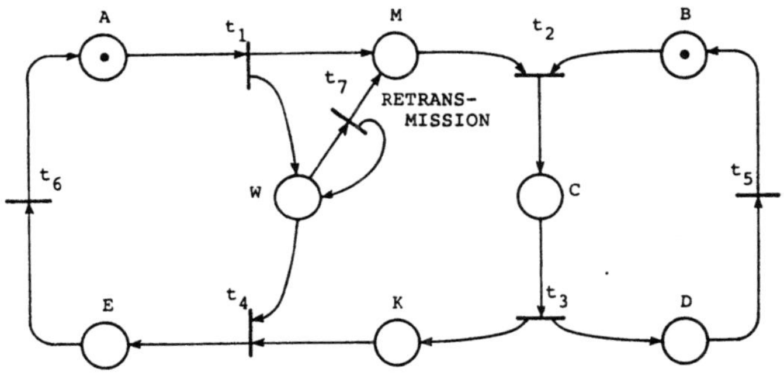

Fig. 5.24. Time Petri net with recovery mechanism

W emits an infinite number of tokens to place M, when there are no assumptions on the firing time of t_7. Transition t_7 cannot fire during regular operation. The sender detects the irregular operation having no acknowledgement within the time-out period, therefore we assume the minimum firing time of transition t_7.

Which transitions should fire in order to help the arrival of the acknowledgement?

$$t_2, t_3 \text{ and } t_4$$

Also the last firing time of the above enumerated transitions must be less than the earliest time of the recovery, namely

$$t_{7\,min} < t_{2\,max} + t_{3\,max} + t_{4\,max}$$

E x a m p l e 5.8 represents the message loss and the origination of a false message introducing the decision function of the E-graph (Fig. 5.25).

If the value of the logical function r is 0, then the token goes into the sink through the X-type transition and the token coming from the source generates a false message.

In **E x a m p l e 5.9** let us represent the connection establishment phase of a transport protocol with a numerical Petri net. The transitions mean the service primitives; the places are control mechanisms and queues. The tokens assigned to the arcs are messages and the token have attributes which are the message parameters (Fig. 5.26).

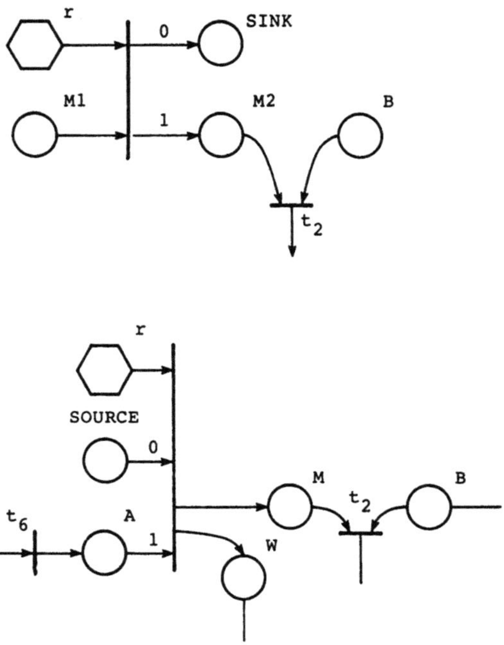

Fig. 5.25. E-graph for message loss

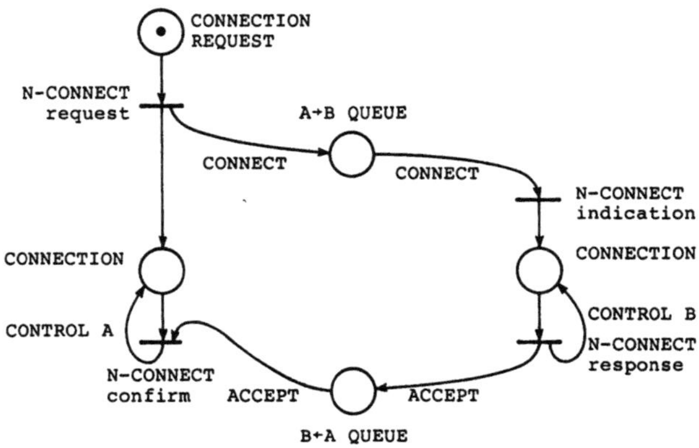

Fig. 5.26. Numerical Petri net to establish connection

6. ALGEBRAS AND FORMAL LANGUAGES

6.1 CALCULUS OF COMMUNICATING SYSTEMS

Introduction

ISO has developed two formal description techniques, LOTOS and ESTELLE, for communication purposes. Both are composed of a mathematical model and language elements. The LOTOS model is based on the Calculus of Communicating Systems (CCS). This section deals with the fundamental notions and basic elements of the CCS and gives an introduction to its syntax, semantics and algebraic structure. Finally an example demonstrates how the CCS can be used for protocol specification.

Basic Ideas

The increasing number of distributed systems made necessary the development of various models of distributed computing. One of these models is the CCS developed by Milner [MILN-80]. Milner's basic idea is "to describe systems in terms of their capability to interact with the environment, therefore systems are described as black boxes having points of interaction with the environment" (Fig. 6.1). The central decision in the CCS is how the interaction can be executed.

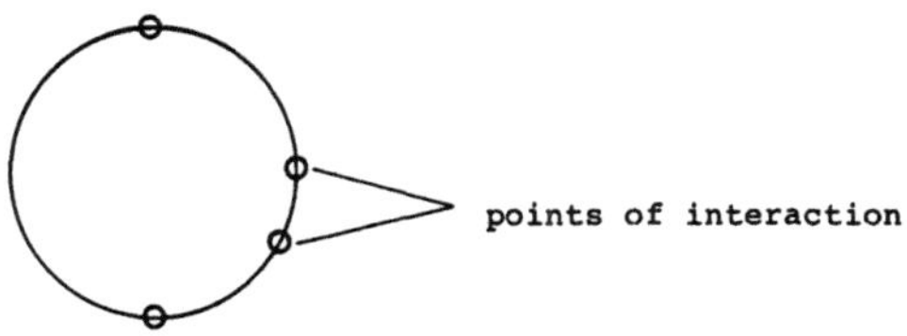

Fig. 6.1. Basic idea of CCS

The execution of the interaction is based on the "rendezvous" principle.

Processes are modelled by agents in the CCS. An agent is an active component which can perform actions. The agent is similar to a finite-state machine and a performed action represents a state transition. The actions are either observable or unobservable. Observation means that an external observer can describe the system from the behaviour of what will be seen or experienced. Unobservable actions are internal actions, generally time-out events.

An observable *interaction* may occur between two or more processes whenever all of them are ready to perform the same action. The interaction point shared by the processes is called a gate. Processes participate in a *"rendezvous"* if they specify the same gate and the information they provide is compatible. If we want to distinguish among the gates, then we can give names or labels to the gates (Fig. 6.2).

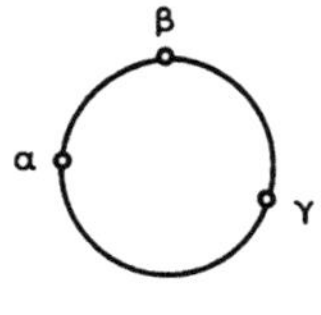

Fig. 6.2. Process with labelled ports

"A behaviour represents the possible activities of an agent which may communicate with its environment." (Milner). Two systems are considered equivalent whenever all external observers cannot recognize any difference between their behaviours. (The behaviours are described in algebraic expressions.)

The "rendezvous" principle means that the communication between two systems is an action which cannot be separated. It appears to be a single action to an observer.

The Syntax and Semantics of CCS

Now we are familiar with the fundamental notions (interaction, behaviour and "rendezvous") and with the basic elements (system, process, agent, action, gate) of CCS.

The syntax and semantics of the CCS is based on *behaviour trees*. There are three kinds of behaviour trees:

— Rigid Synchronization Tree,

— Synchronization Tree,

— Communication Tree.

The Rigid Synchronization Tree (RST) is derived from a state transition graph and has only observable interactions. The Synchronization Tree (ST) is essentially an RST with unobservable, internal interactions. The Communication Tree (CT) is similar to an ST but it has a value-passing possibility.

Figure 6.3 shows a transition tree and its RST pair. The transition tree contains the states and events characterizing the transitions. An observer cannot see the states, only the events are observable. Therefore only arcs of the RST are labelled, the places are "anonymous". More precisely: An RST is a rooted, unordered, finitely branching tree each of whose arcs is labelled by a member of L, where L is the sort of entire label set. The rigidity means that every transition is atomic.

An ST has unobservable actions, their notation is τ. An ST is a rooted, unordered, finitely branching tree each of whose arcs is labelled by a member of $L \cup \{\tau\}$. The arc labelled with τ is a silent transition. Some examples for an ST are demonstrated in Fig. 6.4. The tree is a good way to describe the behaviour of a process. It is easy to follow the temporal order of the interactions of a process with the environment.

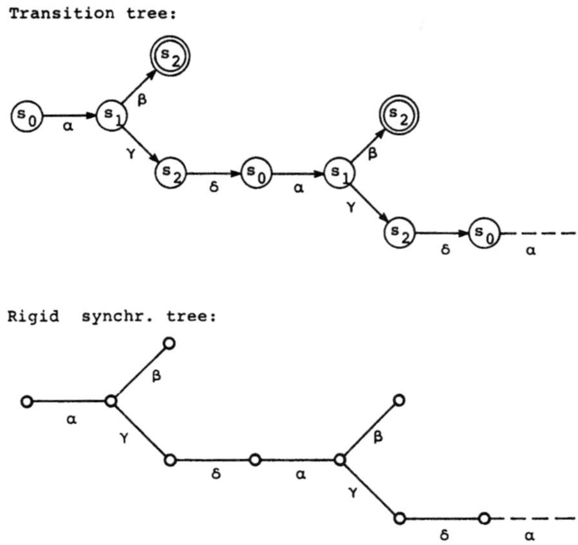

Fig. 6.3. A transition tree and its RST pair

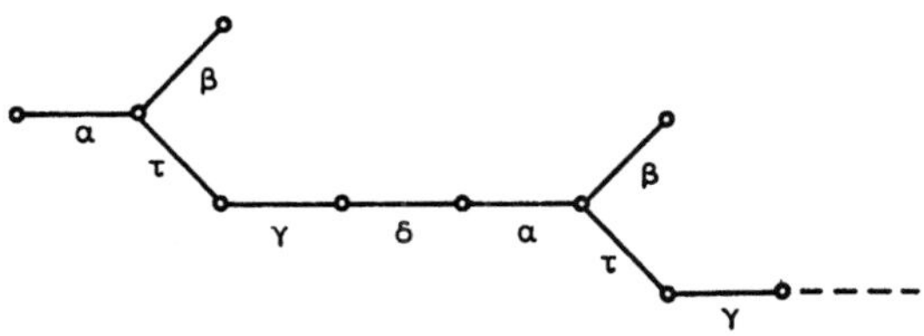

Fig. 6.4. Synchronization tree with the silent transition τ

The behaviour expressions, as we will see later, are represented on the tree.

A CT is a generalized form of an ST permitting value-passing. The value-passing is a parametrized event interaction.

The syntax and semantics of the entire CCS can be found in [MILN-80], we will only deal with the rules of CCS^x in this section. CCS^x is a subset of CCS and this is used in LOTOS. The CCS^x syntax and the CCS^x semantics can be seen in Tables 6.1 and 6.2, respectively [CAR-86].

The fundamental term is "the *behaviour expression*, which describes the dynamic behaviour of the specified system as a process, by implicitly defining all possible sequences of events which can be offered at the gates and are observable from the environment of the system". [BOL-DENI-86]. The syntactic rules define the interpretation of operators between behaviour expressions. A behaviour expression is obtained by applying an operator to other behaviour expressions.

*Table 6.1. Syntax of CCS**

CCS* SYNTAX

Operator	*B*	Remark
Inaction	stop	
Summation	$\Sigma \mathbf{B}$	the binary infix associative version + is also used
Action prefix	$g!e.B1$	*e* is a value expression
	$i.B$	
Composition	$B1 \mid B2$	
Restriction	$B \setminus G$	*G* is a set of gates
Relabeling	$B1[S]$	*S* is a relabeling
Guarding	$e \text{ --> } B1$	*e* is of sort Boolean
Behavior Instantiation	$b(e_1, \cdots, e_n)$	*b* is an identifier

Parentheses must be used to make parsing unambiguous; their use is reduced by the following operator precedences:
(Restriction = Relabeling) > Action > Guarding >
Binary Summation > Composition

*Table 6.2. Semantics of CCS**

CCS* SEMANTICS

Operator	Premiss	Conclusion
Summation	$B \in \mathbf{B},\ B \text{ -}a\text{-> } B'$	$\Sigma \mathbf{B} \text{ -}a\text{-> } B'$
Action prefix	none	$g!e.B \text{ -}ge\text{-> } B$
		$i.B \text{ -}i\text{-> } B$
Composition	$B1 \text{ -}a\text{-> } B1'$	$B1 \mid B2 \text{ -}a\text{-> } B1' \mid B2$
	$B2 \text{ -}a\text{-> } B2'$	$B1 \mid B2 \text{ -}a\text{-> } B1 \mid B2'$
	$B1 \text{ -}a\text{-> } B1',\ B2 \text{ -}a\text{-> } B2'$	$B1 \mid B2 \text{ -}i\text{-> } B1' \mid B2'$
Restriction	$a = gv,\ g \notin G,\ B \text{ -}a\text{-> } B'$	$B \setminus G \text{ -}a\text{-> } B' \setminus G$
Relabeling	$B \text{ -}a\text{-> } B'$	$B[S] \text{ -}S(a)\text{-> } B'[S]$
Guarding	$B \text{ -}a\text{-> } B'$	$\text{TRUE} \text{ -> } B \text{ -}a\text{-> } B'$
Behavior instantiation	$b(x_1{:}s_1, \cdots, x_n{:}s_n) := B_b$	
	$[e_1/x_1, \cdots, e_n/x_n]B_b \text{ -}a\text{-> } B'$	$b(e_1, \cdots, e_n) \text{ -}a\text{-> } B'$

The *inaction* has a nullary operator called *stop*. No axiom or inference rule is associated with it. It denotes a process which is unable to perform any action and cannot interact with any other process.

The *summation* is a non-deterministic choice and combines the possible activities of the behaviours, allowing one or the other to proceed. **B** is the set of behaviour expressions.

The *action prefix* has two different forms for observable and unobservable actions. The observable action in Table 6.1 appears at g, where $g \in G$. G is any set of gates. The value expression e can be a single or multiple value offer. If it is a multiple value offer, the action prefix is

$$g\,!\,x:t\,.\,B\,,$$

where x is a variable declared of sort t. The action prefix operator i is for unobservable cases. If we apply operator i to a behaviour expression B, the resulting behaviour performs an unobservable action and transforms into B.

The *composition* allows behaviours to proceed concurrently with possible intercommunication between complementary labelled ports. This implies that B1 is ready to wait in the absence of alternative actions until its partner B2 offers the same action.

The *restriction* operator transforms the action occurring at the selected gate into an unobservable action.

The *relabelling* function S is specified by $g_1...g_i...g_n/h_1...h_i...h_n$, where g_i are the gates whose names are to be changed into h_i by S.

The *guarding* is a logical condition. The behaviour B1 can occur when e is true.

The *behaviour instantiation* determines behaviour B depending on the behaviour identifier b. Any identifier produces different behaviour expressions having different value expressions.

The rules of precedence are as follows:

(Restriction, Relabelling) > Action > Guarding > Summation > Composition.

The CCS^x operators are shown in Fig. 6.5, showing how behaviour expressions are constructed.

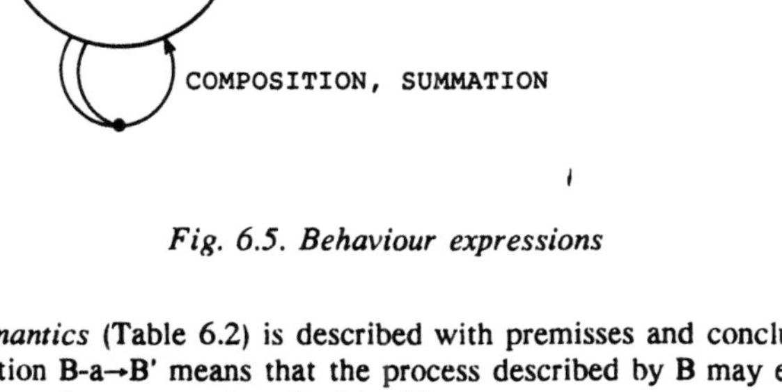

Fig. 6.5. Behaviour expressions

CCS^x *semantics* (Table 6.2) is described with premisses and conclusions.

The notation B-a→B' means that the process described by B may execute action a, and it becomes the process described by B'.

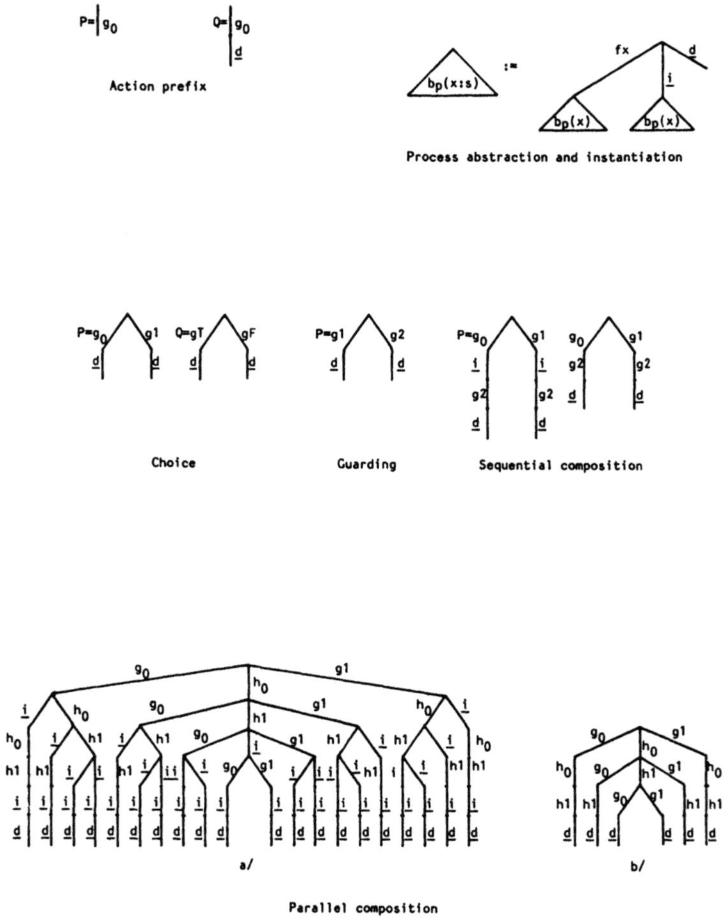

Fig. 6.6. Behaviour trees

The behaviour trees are shown in Fig. 6.6. Every tree has a starting node representing the starting state of the process. The edges leaving this node represent the paths to the next node. The edges are labelled with the actions to be executed.

Inaction: The first operator *stop* is missing from Table 6.2 because no action belongs to it. Its behaviour tree is a single node.

Summation: The behaviour of $\Sigma \mathbf{B}$ will be the behaviour of any of the expressions $B \in \mathbf{B}$. The behaviour tree is described by any branch or subtree of **B**.

Action prefix: The action is either g ! e or i. An observable action prefix means an offered value e at gate g, an unobservable one is an internal action i. The edge of the behaviour tree is labelled with the action prefix. The "!" symbol and the "?" symbol indicate output transition and input transition, respectively.

Composition: The construct B1 | B2 describes concurrent behaviours of B1 and B2 with communication through identical actions. The composition can add temporal ordering constraints to an already specified behaviour.

Restriction: The behaviour of B \ G can hide G from the environment of B. Any subtree starting with a G-labelled edge is pruned on the behaviour tree. The restriction operation simply replaces the G-labelled edges with i-labelled ones.

Relabelling: The behaviour of B[S] is the same as B, but the actions are relabelled by S. Relabelling allows one to consider different instances B[S] of a behaviour expression B, differing only in the gate names. The original and the relabelled tree have the same structure, only on the relabelled tree the gates are renamed.

Guarding: The behaviour of a guarded process is B if the value expression e is true and is *stop* if e is false. The behaviour tree is either the tree of B or the node *stop*.

Instantiation: Given the behaviour abstraction b $(X_1 = S_1, \ldots, X_n = S_n) := B_b$. It is an abstraction because the behaviour is defined in terms of gates and variables. The behaviour instantiation means replacing the above terms with actual values. A recursive instantiation can be executed either directly or through a chain of nested instantiations.

Equivalences

Milner defines some type of equivalences (Fig. 6.7). The algebraic laws are formulated with these equivalences. The equivalences start with direct equivalence and terminate with the observation equivalences in a broad range, from the identity relation to the universal relation.

Direct equivalence: Two behaviours are directly equivalent if they are interchangeable without affecting the behaviour derivation of the whole. Each behaviour has the same result.

Strong *congruence:* Two behaviours are strongly congruent if the derivative of one behaviour is the same for some values of the other behaviour, and the opposite is also true. A derivative of a behaviour is a sequence of behaviours performing a sequence of experiments.

Observation congruence: Two behaviours are observation-congruent if their derivatives cannot be distinguished by observation in any context.

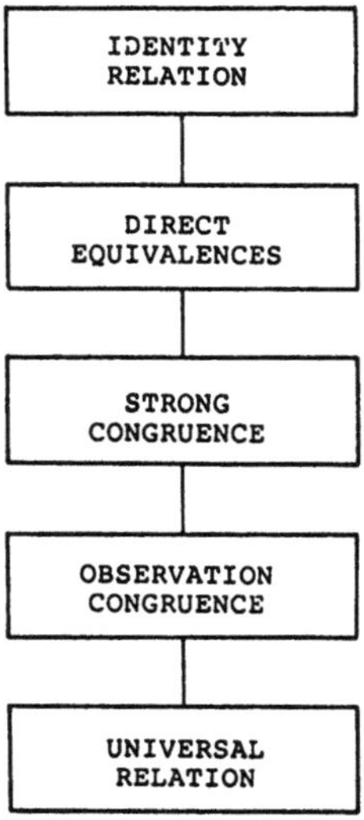

Fig. 6.7. Milner equivalences

Observation equivalence: Two behaviours are observation-equivalent if they cannot be distinguished by any finite observation. It allows abstraction from unobservable actions.

The equivalences become weaker going from direct equivalence to observation equivalence. The two equivalences are valid for behaviours. The two congruences for the derivatives perform a sequence of atomic experiments.

Example

A simple protocol specification demonstrates the power of the CCS in this subsection. The example is based on [PARGU-84].

We have a very simple protocol (Fig. 6.8) which sends and receives messages. The communicating medium is also modelled. The medium can lose messages. In this case a time-out "to" signals this to the sender and the sender then retransmits the message. It should be emphasized that the CCS does not provide a means of expressing timing-constraints, e.g. when a time-out will occur. A non-deterministic choice is specified between time-out and acknowledgement in our example. The receiver sends the acknowledgement directly to the sender.

The components of the protocol are the following in the CCS-interpretation. There are three processes: the Sender (S), the Receiver (R) and the Medium (M). The protocol has two external gates: *send* and *rec* and four internal gates *in, out, to* and *ack*. The external gates ensure a connection with the environment. The internal gates are hidden from the environment. Figure 6.8 explains the interpretation of the different gates.

The specification of the Sender is

S = send? . in! . (to? . S + ack? . S) .

The specification of the Receiver is

R : out? . rec! . ack! . R

The specification of the Medium is

M = in? . (out! . M + i.to! . M) ,

where
i is an internal action.

The protocol specification is the composition of the three processes restricting the internal gates:

P = (S | M | R) \ {out, in, to, ack} .

The specification can be decomposed into detailed elements, but this simple specification also shows well the expressive power and the steps of communication.

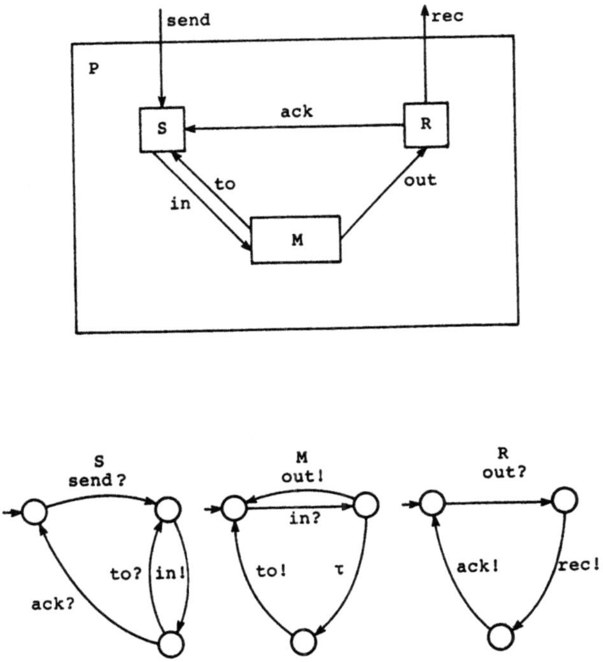

Fig. 6.8. Simple protocol specified by CCS

6.2 A SPECIFICATION LANGUAGE FOR ABSTRACT DATA TYPES

Introduction

A Formal Description Technique cannot be complete without specifying the data. Programming languages use concrete data types which are implementation dependent. We need a formal method to describe generally the data. We create a mathematical model of the concrete data type which is independent of data structures, implementation choice and data codes. The abstract data type is this mathematical model.

ACT ONE is a specification language for Abstract Data Types (ADT) and it is integrated into LOTOS (see section 7.4).

ACT ONE was developed by the ACT-group at the Technical University of Berlin [Ma-85]. ACT ONE is an algebraic specification method.

The rest of this section deals with ADTs and the language concepts of ACT ONE, finally some example ADTs are presented.

Abstract Data Types

An ADT is a high-level data abstraction, because the concrete data structures are not fixed and the temporal execution of the operations is not defined. There are different methods for specifying ADTs, one group of them is based on algebraic methods.

A specification of an ADT contains two parts:

— the type and range of the data,

— the set of possible operations.

The first part determines the syntax describing the names, definitions, ranges and the validity of operations. The second part specifies the semantics of the ADT giving the axioms. Essentially the abstract data types are algebras [GUT-77, GOTZ-85].

The specification of an ADT is a triplet:

$$SPEC = (S, OP, E)$$

where

S is the set of sorts,

OP is the set of operation symbols,

E is the set of equations.

We can, for instance, specify the ADT for simplified Boolean values having only two operations, NOT and AND.

The set of sorts is

$$S = (boole) .$$

The set of operations symbols is

$$OP_1 = (TRUE, FALSE) ,$$

$$OP_2 = (NOT, AND) .$$

The set of equations is

NOT(TRUE) = FALSE

NOT(NOT (B)) = B

AND(B, TRUE) = B

AND(FALSE, B) = FALSE

AND(B, FALSE) = FALSE .

Generally, the mapping rules are added to the operation symbols:

NOT : boole → boole ,

AND : boole → boole .

Summarizing, the power of ADTs is in organizing the system. We will discuss other examples in the next sections, but first we will learn the notation of ACT ONE and then use this in the examples.

Language Concepts of ACT ONE

ACT ONE is a specification language for ADTs and deals with an algebraic specification method for writing parametrized and unparametrized specifications based on the same mechanism [EHR-86].

ACT ONE has the following features:

- Modularization of specification which allows the reference to already existing specifications in a library.
- Combination of specifications.
- Renaming of specifications.
- Parametrizing of specifications.
- Actualization of parametrized specifications.
- Extension of specifications with operations and sorts.

The above features will be demonstrated at the end of this section.

If we want to define an ADT, we define the names of the data carriers and the operations. The definition must also include the domains and ranges of the operations. The names of the data carriers are referred to as sorts. The definition of these sorts and operations is called the signature. The signature defines the syntax of a data type:

$$sign_1 = [sorts][opns] .$$

The names of the sorts and the operations must correspond with the data carriers and operations of a distinguished algebra — the semantics of the specifications. All the elements of the data carriers that can be produced by repeated calls of the operation symbol call are called a term.

For example, if we have a specification of the natural numbers represented by the sort "nat", a constant "0: > nat" and a unary operation "succ: nat" then we can produce the following nat-terms:

0, succ(0), succ(succ(0)), ...

Each of these nat-terms can be interpreted as one element of the algebra of the natural numbers.

All elements of an algebra can be represented by the concept of the signature. For instance, we can introduce additional operation symbols.

The plus operator is

$$+: nat, nat \rightarrow nat$$

We shall enlarge our concept of the signature with equations to get our algebra:

$$sign_2 = [eqns] .$$

The equations contain variables and rules. The informal introduction of the plus operator can now be formalized by using the concept of equations. The new operator must be consistent with the already available operators "0" and "succ":

EQNS FOR ALL X, Y IN nat: ,

X + 0 = X ,

X + succ(Y) = succ(X + Y) .

The first equation expresses the semantics of the plus operator, if it is combined with the constant "0". Also, the "succ" operator was introduced to have access to the natural number with the plus operator, since the result of its application can be represented by the "succ" operator.

Abstract Data Types of ACT ONE

The specification of an ADT in ACT ONE begins with the keyword TYPE and ends with ENDTYPE. The detailed specification is between the keywords. The entire structure of the ADT can be seen in Fig. 6.9.

There are atomic and enriched (complex) specifications. The atomic specification can be

— simple,
— formal,
— parametrized.

The complex specification can be

— a combination,
— a modularization,
— a renaming,
— an actualization.

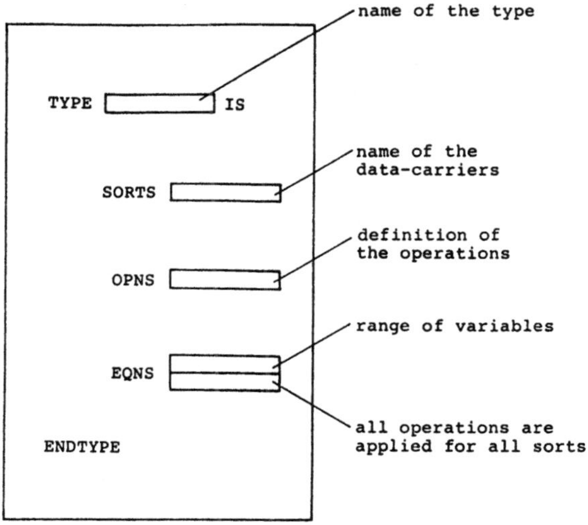

Fig. 6.9. Specification of an abstract data type in ACT ONE

The complex or enriched specification is an important feature of ACT ONE. The essence of the complexity is that

- simple specifications may be combined stepwise (COMBINATION),
- large specifications can be split into smaller parts (MODULARIZATION),
- the whole specification may get a new name (RENAMING) ,
- the parameter values can be actualized in a specification (ACTUALIZATION).

Atomic Specification

The atomic specification is a single unit. The general form of a simple specification is the following:

```
type   name of specification   is
       sorts   list of sorts
       opns    list of operations
       eqns    list of equations
end of type
```

The atomic specifications contain no definitions of variables.

Two examples demonstrate the simple specification describing the basic Boolean values and the integers.

The specification of the Boolean values is the following:

```
type   bool 0    is
   sorts
               bool
   opns
               TRUE, FALSE: → bool
end of type
```

The ADT of integers uses the operators successor (SUCC) and predecessor (PRED) representing the next and previous component respectively.

```
type    int 0    is
   sorts
               int
   opns
               0: → int
               SUCC, PRED: int → int
   eqns
               SUCC (PRED(x)) = x
               PRED (SUCC(x)) = x
end of type
```

The general form of a formal specification with name formalspec is denoted:

```
type    formalspec    is
   formal sorts    sortlist
   formal opns     oplist
   formal eqns     eqlist
endtype
```

The application of the formal specification will be demonstrated in the parametrized specification.

The parametrized specification is the third atomic specification. Its general form is:

```
type  pspec  is
    formal sorts   sortlist
    formal opns    oplist
    formal eqns    eqlist
    sort:    sortlist 1
    opns:    oplist 1
    eqns:    eqlist 1
endtype
```

Typical examples of parametrized specification are arrays, stacks, queues.

As an example of parametrized specification we introduce the specification of a string. The "empty string" is distinguished among the strings. There are four string operations:

— concatenation (CONCAT),
— left-side addition (LADD),
— right-side addition (RADD) and
— making a string from data (MAKE).

The definition contains the parameter name in the first line:

```
type  string (data)  is
    formal sorts   data
    sorts   string
    opns    EMPTY: string
            MAKE: data → string
            CONCAT: string string → string
            LADD: data string → string
            RADD: string data → string
```

```
    e q n s   o f   s o r t   string
              f o r   a l l   d   i n   data:
              s, s1, s2, s3   i n   string:
                 CONCAT (s, EMPTY) = s
                 CONCAT (EMPTY, s) = s
                 CONCAT   CONCAT ((s1, s2), s3) =
                     = CONCAT (s1, CONCAT (s2, s3))
                 LADD (d,s) = CONCAT (MAKE(d), s)
                 RADD (s,d) = CONCAT (s, MAKE(d))
e n d t y p e
```

Complex Specification

If we execute some operations on the atomic specification we get a new, complex specification.

The combination is the union of many finite specifications. The general form of a combination is

```
t y p e   name   i s
    name-1  a n d  name-2  a n d ... a n d  name-n
    f o r m a l   s o r t s
    f o r m a l   o p n s
    f o r m a l   e q n s
    s o r t s
    o p n s
    e q n s
e n d t y p e
```

where every name should be specified as

```
type    name-i    is

        [          ]

endtype
```

As an example we shall define the length of a string assuming that the data-types for natural and string (data) are known:

```
type      string-1 (data-1)    is
          string (data)    and    nat    and
    opns      LENGTH: string → nat
    eqns    of    sort    nat
          for    all    d    in    data:
          s    in    string
          LENGTH (LADD(d,s)) = SUCC (LENGTH(s))
          LENGTH (EMPTY) = 0
endtype
```

Modularization helps to construct complex specifications from simple specifications.

Renaming gives a new name for a defined data-type. The general form of renaming is the following:

```
type      new name    is
          old name    renamed    by
          sort    names    [          ]
          op    names      [          ]
endtype
```

The blank boxes contain

```
new-sort-i    for    old-sort-i
new-opns-i    for    old-opns-i
```

For instance

```
type    connection   is
        queue   renamed   by
        sort   names   channel   for
                       queue
                       objects   for
                       data

        op   names     send   for
                       add
                       receive   for
                       first
endtype
```

Finally, actualization is a central operation in ACT ONE. Actualization is the substitution of the formal parameters with concrete parameters

```
type   new spec   is
   param-spec-1   actualized   by
   param-spec-2   and ... and   param-spec-n
   using   sort   names   [          ]
           op   names     [          ]
endtype
```

where in the blank boxes are

```
concrete-sort-i   for   formal-sort-i,
concrete-opns-i   for   formal-opns-i,
```

respectively.

Many examples of ADTs can be found in [EHR-83].

It should be noted that the specification in the original ACT ONE is between *def* and *end of def* instead of *type* and *endtype*.

6.3 FORMAL LANGUAGES

Introduction to Formal Languages

The human, natural language is an important tool of human communication. The formal languages were first developed to model the natural languages. Later the formal languages became more general and they were used for system description.

A language (i.e. a formal language) is a set of sentences and a sentence is a finite sequence, or string, of elements, each of which is drawn from a finite vocabulary. For example, given that the vocabulary V of language L is the two-member set {a,b}, L might have as its members not only a, b, ab and ba, but also aa, aba, bba, etc. [LYO-82].

The *length* of a string is determined by the number of elements of which it is composed. For example, given that V = {a,b}, aa is a string of length 2 over V, bab is a string of length 3 over V. Strings may be finite or infinite in length, but sentences are defined to be of finite length. Strings are formed by means of the binary operation of concatenation. The shortest string has only a single element, its length is 1. The empty string or null string has no elements and its length is zero. The set of all strings over V is denoted as V^*.

A language can be defined by a particular grammar as the set of all the sentences it generates. The *grammar* is an ordered quadruple.

$$G = (V_N, V_T, F, S) ,$$

where

- V_N is the set of non-terminal symbols (finite alphabet):
- V_T is the set of terminal symbols (finite alphabet);
- F is the rule system of the grammar,
- S is the starting component of the generation.

Terminal elements are those which actually occur in sentences. All other symbols that are employed in the formulation of grammatical rules may be described as auxiliary or *non-terminal* elements.

The terminals and non-terminals create a disjunct set

$$V_T \cap V_N = 0 .$$

The *starting component* S belongs to the set of non-terminals

$$S \leftarrow V_N .$$

The *grammatical rule system* F contains the rules of substitution composed of the production rules. The production rule has the form

$$P \rightarrow Q .$$

$P,Q \in (V_N \cup V_T)^*$ and F is the finite set of the ordered pairs (P,Q). There is one non-terminal among the elements of P by all means.

Classification of Grammars

Chomsky classified the grammars according to the production rules. He introduced four types:

— type 0: phrase structure grammars;
— type 1: context sensitive phrase structure grammars;
— type 2: context free phrase structure grammars;
— type 3: regular grammars.

The production rule of type 0 is:

$$X \rightarrow Y ,$$

where

$X,Y \in (V_N \cup V_T)$.

The rewriting on the right side of the derivation is unrestricted.

The phrase structure is comparable with the notion of bracketing in mathematics or symbolic logic.

The production rule of type 1 is:

$$X \rightarrow Y ,$$

where

$l(Y) \geq l(X)$.

The derivation or substitution depends on the context, e.g.

$$Q_1XQ_2 \rightarrow Q_1PQ_2$$

The non-terminal **X** will be substituted with the terminal **P**,if and only if the direct context of **X** is Q_1 and Q_2.

The production rule of type 2 is:

$$X \rightarrow P ,$$

where

$X \in V_N$

$P \in (V_N \cup V_T)^*$

This grammar generates context-free sentences.

The context-sensitive and the context-free grammars are both phrase structure grammars. They have the property that they not only generate strings of terminal symbols, but they assign to each such terminal string a particular kind of structural description - an analysis of the terminal string in terms of a labelled bracketing. To assign such a structural description to the terminal strings that they generate is the function of the auxiliary or non-terminal symbols.

The grammars of type 3 are the regular grammars. The production rules derive from non-terminal X either a terminal or the concatenation of a non-terminal and a terminal:

$$X \rightarrow P \text{ or } X \rightarrow PY$$

where

$X,Y \in V_N$
$P \in V_T^*$

The regular grammars are called finite-state grammars, too. These finite-state grammars are used on the protocol field and therefore some remarks will be given.

This grammar is capable of generating an infinite set of sentences by means of a finite number of recursive rules operating upon a finite vocabulary. These are based on the view that sentences are generated by means of a series of choices made "from left to right": that is to say, after the first, or leftmost, element has been selected, every subsequent choice is determined by the immediately preceding elements.

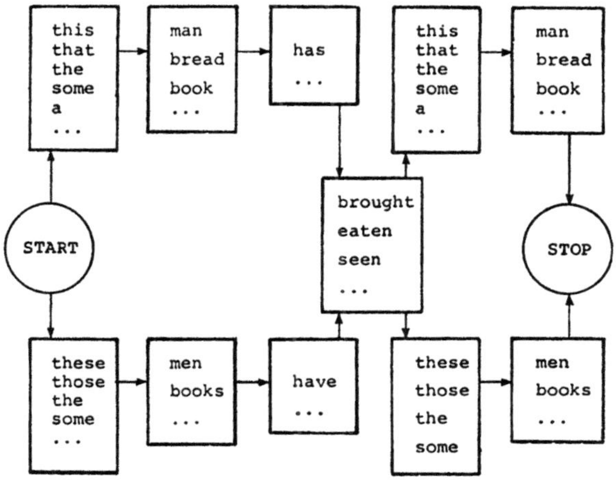

Fig. 6.10. State diagram for English sentences

Figures 6.10 and 6.11 show the state diagrams for English sentences and for simple protocols. The diagrams may be interpreted as follows. We can think of the grammar as a machine, which moves through a finite number of internal states as it passes from the initial state (START) to the final state (STOP) in the generation of sentences. When it has produced a word (from the set of words given as possible for that state) the grammar then switches to a new state. Any sequence of words that can

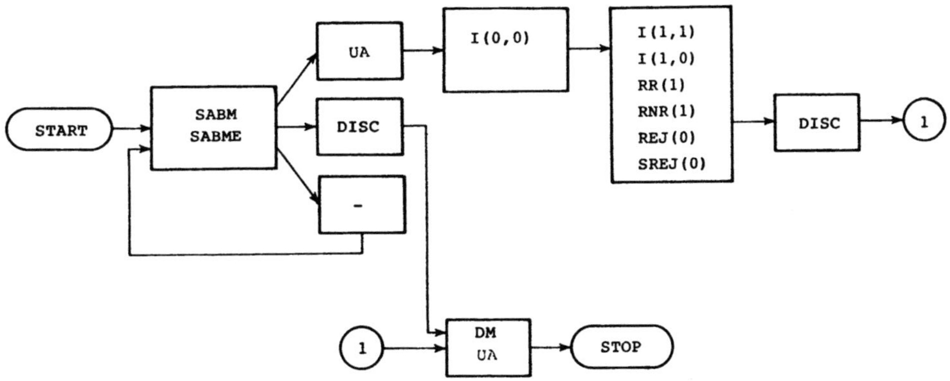

Fig. 6.11. State diagram for a Simple protocol

be generated in this way is thereby defined as grammatically correct. The grammar illustrated in Fig. 6.10 generates a finite number of sentences, the other grammar in Fig. 6.11 generates a dialogue between two entities.

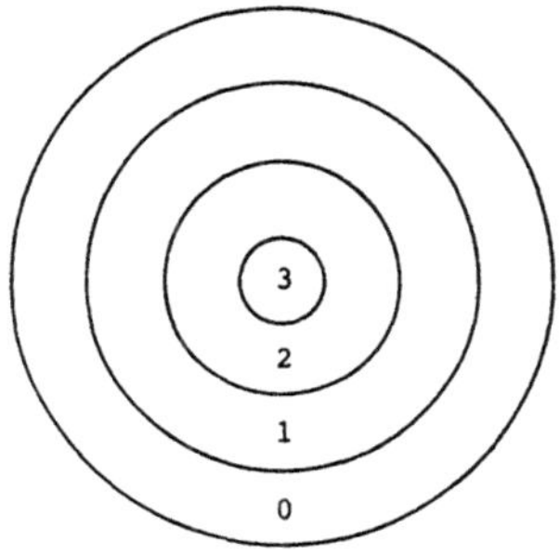

Fig. 6.12. Embedded grammars

The grammars are embedded (Fig. 6.12). They have a hierarchical classification. The languages generated by grammars of type 4 are properly included as a subclass of the languages generated by grammars of type 3; the languages generated by grammars of type 3 are properly included in the languages generated by grammars of type 2; and in general, the languages generated by grammars of type i are also generated by grammars of type i-1, but not conversely [CERF-72], [CHOM-75], [GIN-66], [GOS-71], [REV-79].

7. SPECIFICATION LANGUAGES FOR COMMUNICATION

7.1 SPECIFICATION AND DESCRIPTION LANGUAGE (SDL)

Overview of SDL

The increasing number of functions performed by the new computer controlled telecommunication systems stimulated the CCITT to develop and standardize languages for system specification, system description and for writing and reading unified documents. The CCITT developed three languages, each has another main application field. The CCITT Software languages are the following:

— SDL (Specification and Description Language)
  • to facilitate the specification of functional requirements
  • to aid the high level description of the functional behaviour of implemented systems.

— CHILL (CCITT High Level Language)
  • to use as a programming language for stored program controlled (SPC) switching systems.

— MML (Man Machine Language)
  • to facilitate operations, maintenance, installation and testing by providing consistent input/output interface
  • to use in switching exchanges and in operations and maintenance centres.

Though each of the CCITT standard languages has a well-defined field of application, the three may be considered as a whole, covering the various aspects of the software in telecommunication. The SDL is used in the protocol field, therefore further on we deal with SDL.

It was in 1973 that the CCITT enabled the development of SDL to start. At first a model for representing computerized telecommunication functions was discussed, based on intercommunicating extended finite-state automata and next a set of basic concepts and their representation in a graphic form was approved. In 1976 the first set of Z.100 Recommendations was ready to be published and the language was officially named SDL as the functional Specification and Description Language of CCITT.

A new set of Recommendations was Prepared and approved in 1980 and then SDL started to be used in several organizations. Later the Recommendations were extended providing more features for system structuring, refinement and data.

The SDL Recommendations are composed of five parts

— General definition of SDL: Z.100,
— Basic concepts: Z.101,
— Structuring concepts: Z.102,
— Refinement: Z.103,
— Data: Z.104.

This section familiarizes us with the SDL and the structure of this section follows the structure of the above-mentioned Recommendations [SARA-84].

General Definition of SDL

The specification and description in SDL is for functional representation of a system. A *specification* of a system is the description of the required behaviour and a *description* of a system is the description of the actual behaviour of the system. SDL describes the behaviour as a "stimuli-response" pair. The approach is based on the concepts of the extended finite-state machine. The interpretation of this approach will be explained when dealing with the process definition.

The behaviour of a system described in SDL is the sequence of responses to any given sequence of stimuli as seen from the outside of the system.

A complex system is divided into parts, each of these parts is called a block. The process describes the functional behaviour of a block.

A semantic model and an abstract syntax help us to define the language. The semantic model is the set of rules relating one element to the others. The abstract syntax gives a precise meaning to each element and rule. The abstract syntax is transformed to concrete syntax. There are three concrete syntaxes in SDL:

— graphic syntax (SDL/GR)
  it represents the functional behaviour of a system with a set of graphic symbols and defines the rules connecting the symbols,

— program-like syntax (SDL/PR)
  the syntax is very similar to the syntax of the Pascal programming language, the symbols are reserved keywords,

— pictorial element syntax (SDL/PE)
  it is an extension of the SDL/GR, the pictorial elements contain more information than in SDL/GR.

Basic Concepts

A *system* is composed of a non-null set of blocks connected to each other and to the system environment by means of channels (Fig. 7.1). Channels are unidirectional. A *block* should have at least one incoming channel and at least one outgoing channel. The incoming and outgoing channels are the only relationship of a block with its environment. No other exchange of information is possible. A block is defined when the processes modelling its behaviour, its data, and its boundary are defined. A unique name is associated with a block.

The *process* is the model of the functional behaviour of a block. A process is defined when all the sequences of operations following any sequence of stimuli are defined. A unique name is assigned to each process.

Each process's behaviour is represented by an extended finite-state machine model. A process is composed of conditions awaiting stimuli and executing response-operations. The conditions depend on the states and the stimuli and the response-operations — on the events of the state transition graphs. Each process has access to local data in addition to its state value and the process can test and change this local data during its transition between states, therefore this model is called extended finite state-machine model (in a classical finite-state machine model, the only item of data visible is the state number).

The *signals* are the means by which information can be moved from one process to another. Information is transferred between the environment and the system, and within the system. Signals represent the stimuli that activate processes. Therefore signals have a twofold aspect, namely information transfer and process activation. Processes model the expected sequences of information transfer and the computation following each activation.

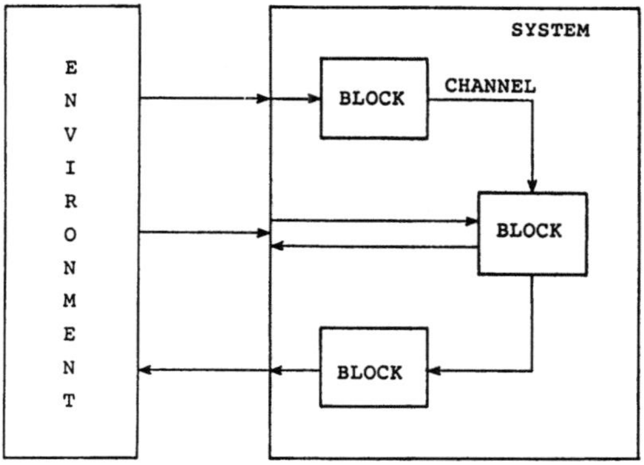

Fig. 7.1. System definition in SDL

The signals are conveyed on the *channels*. The channel is unidirectional: this means that signals flow in one direction only. In the case where two blocks are exchanging signals in both directions a pair of channels is required. The modelling of the channel as unidirectional simplifies the model.

Each channel has an originating port and a terminating port. A port may be associated with more than one channel.

The channel is defined as a transportation medium which does not perform any operation on the signals transported.

Blocks, processes, signals and channels are actualized by their *instances*. An instance is characterized by its data. The actual dynamic behaviour of a system and its complete structure can only be fully understood considering instances.

The definition of any basic elements of SDL starts with a name. The definition of an element requires that other elements in lower hierarchical level should be defined, too [BEL-SAR-86], [STE-86].

SDL Process Diagram

The process is defined by a process graph and may contain associated data. The process graph is the method of representing the process behaviour. A process, at any given time, is either in a *state* or in a *transition* between its states.

A process is in a transition in consequence of the recognition of an input and follows the transition up to a point where another state is reached. At that point the process waits for the recognition of another input.

This extended finite-state automaton has a high graphic illustrative power and is easy-to-learn and easy-to-use.

Table 7.1. Graphic representation in SDL

Graphic representation	Name	Interpretation of the components	Interpretation of the graphic representation
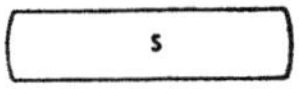	STATE	S state name	The concept of state is used similarly as in finite state machines
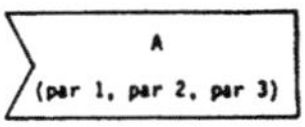	INPUT	A signal name par 1, par 2, par 3 } parameters with local variable names	The input is an arriving signal recognised by the process
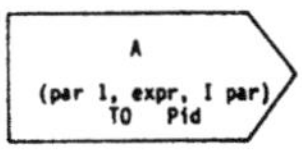	OUTPUT	A signal name par 1 parameter expr expressions imported/ I par /received values keyword destination TO (process Pid instance identity)	The output is an action inside a transition generating the next signal
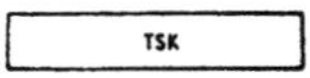	TASK	TSK task name	The task can be any action inside a transition expect decision and outputs
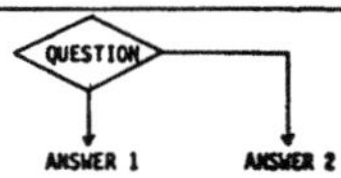	DECISION		The decision is an action inside a transition putting a question. The transitions can be continued in different directions depending on the answer
	START		The start action if any is the first action of a process and is followed by a transition
	STOP		A STOP terminates the process instance
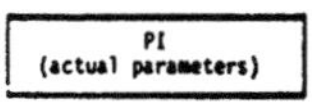	CREATE	PI process instence	The "creation" action causes the creation of the indicated process instence

The process graph contains a set of nodes connected by directed arcs. Each node may take the form of one of the following types:

— state node,
— input node,
— output node,
— task node,

— decision node,
— start node,
— stop node,
— create node.

The graphic symbols of the nodes and their main features can be seen in Table 7.1.

The *process diagram* is the concrete representation of the process graph. Drawing rules for process diagrams:

— start with state,
— draw top to bottom, left to right,
— avoid line crossings,
— uniform symbol size,
— highlight major aspects.

A process diagram usually consists of many states, inputs and transitions. The diagram can be split to avoid too extensive figures. We can break the drawing with a state and resume it on another page starting with the same state (Fig. 7.2).

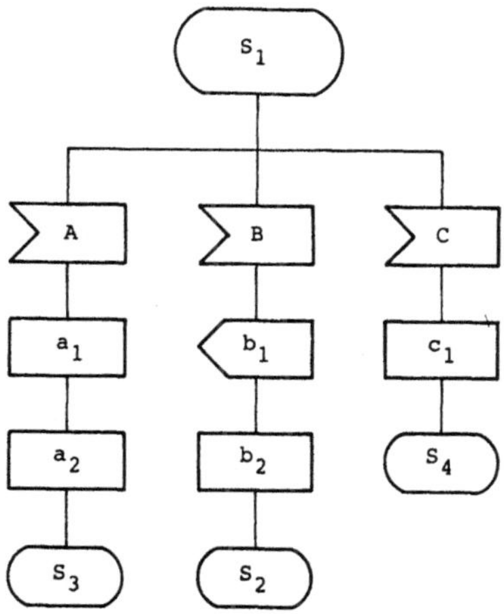

Fig. 7.2. Process diagram

STRUCTURING CONCEPTS AND REFINEMENT

From the Basic Concepts to the Structuring Concepts

The basic concepts of SDL can be summarized briefly as follows:

> "An SDL system has a set of blocks. Blocks are connected to each other and to the environment by channels Within each block there are one or more processes These processes communicate with one another by sig nals and are assumed to execute concurrently"
>
> SDL Recommendation

These basic concepts are suitable for specifying relatively small systems which may be understood and handled at a single level of blocks. When a larger, or complex system needs to be specified, it is necessary to partition the system specification into manageable units, which may be handled and understood independently.

Partitioning does not affect the static interface of a unit. In addition to partitioning, there is also a need to add new details to the behaviour of a system. This is denoted by the term refinement.

So the structuring concepts of SDL have two central notions: partitioning and refinement.

Partitioning is the subdivision of a part of a system into smaller parts whose global behaviour is equivalent to the unpartitioned part. It can be applied to blocks, to channels, and to processes. The blocks will be structured in new subblocks, channels and subchannels. The channel will be structured in blocks, new channels and subchannels. The processes will be structured in services.

Refinement is the addition of new details to the system functionalities. The refinement of a system causes an enrichment in its behaviour as more kinds of signals and information can be handled.

Partitioning

SDL starts with a high level view of a system representation: a system consists of a set of blocks connected with channels, and the blocks contain processes. The partitioning breaks the system into manageable pieces. The blocks, the channels and the processes can be partitioned.

The block partitioning can be represented by the block tree diagram (Fig. 7.3) where we have at the top the partitioned block connected to the subblocks. The block partitioning causes the appearance of "new-channels". The "new-channels" help to realize the information exchange between the subblocks.

The partitioning of a block may lead to the partitioning of the incoming and/or outgoing channels. The partitioning of a channel generates a set of subchannels that globally contain the same information as the partitioned channel.

Two ways of channel partitioning can be seen in Fig. 7.4.

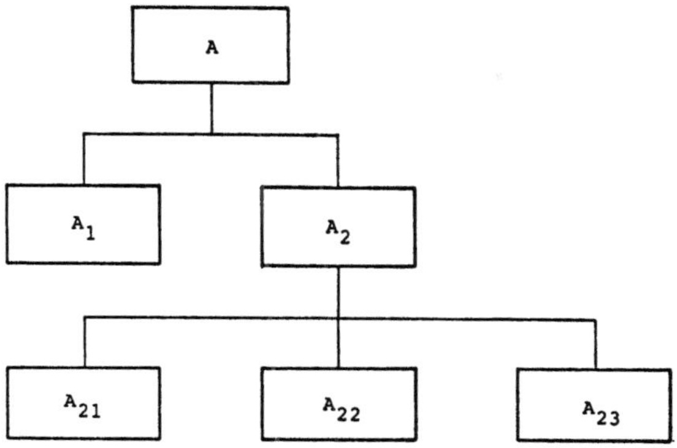

Fig. 7.3. Block tree diagram

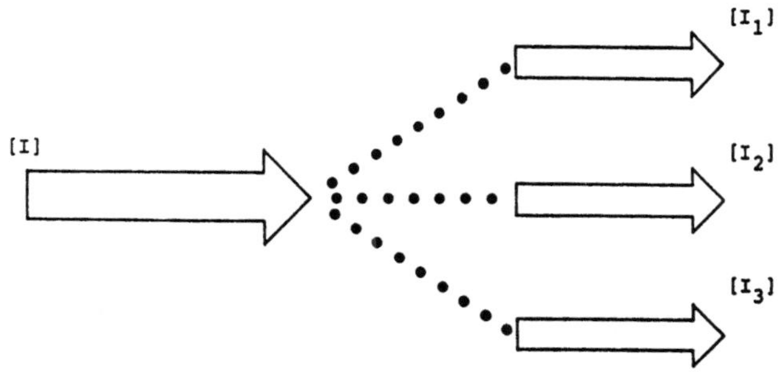

$[I] = [I_1] + [I_2] + [I_3]$

Fig. 7.4. Channel partitioning

The third possibility for decomposition is the process partitioning. A process represents the behaviour of any block. A process can be split into subprocesses (Fig. 7.5). A process receives a set of information in the form of signals and releases information by sending signals. As the effect of the decomposition of the process, the incoming information is split so that each input signal can be received by one and only one subprocess. This rule does not apply to the outgoing information where we may have several subprocesses sending the same signal.

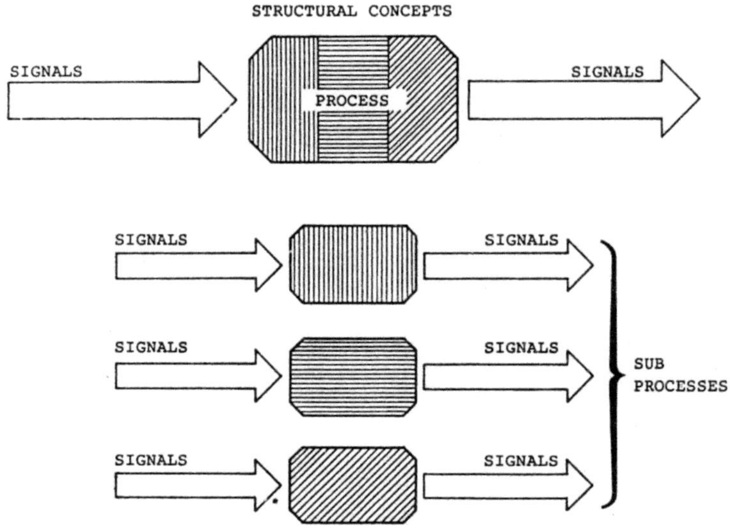

Fig. 7.5. Structural concept

Refinement

The refinement concepts extend the facilities of the SDL in the representation of systems. The complexity of diagrams is enriched by using

— procedures,
— macros,
— continuous signals,
— the activation of transitions according to some conditions.

The *procedure* concept represents a mean both of reducing the complexity of an SDL diagram, and of reflecting the actual system structure. At the same time the use of procedures gives the user the possibility of a step by step refinement method in the definition of a process.

Basically a procedure is a process that is called directly from another process. Its activation is made at the call time, at a certain point the interpretation of the procedures terminates and the process continues at the node following the one which called the procedure. Any SDL node may include a procedure, with the exception of the stop node.

The *macro* construct may be used:

— to draw repetitive parts only once,
— to avoid drawing a complex part on a main document,
— to highlight a certain part with a meaningful name.

It is important to emphasize that a macro has no semantics of its own but derives it from the context.

The procedure and the macro are compared in Table 7.2.

Table 7.2. Comparison of the procedure and macro concepts

PROCEDURE	MACRO
- can only be applied to process	- can be applied to any document
- local visibility	- no visibility of its own
- has parameter	- no parameter
- can be called from any place where a task is legal	- can be called everywhere
- interpreted at processing time	- interpreted before processing time

The construct of a *continuous signal* is a powerful shorthand notation that allows us to initiate a transition depending on the condition of a variable visible to the process. A continuous signal can be expressed by a certain set of basic SDL constructs. A continuous signal is depicted in Fig. 7.6. The condition is always a Boolean expression. If the condition is true, the transition is activated. If the condition is false, the process instance waits. Therefore the activation of transitions is strongly related to continuous signals.

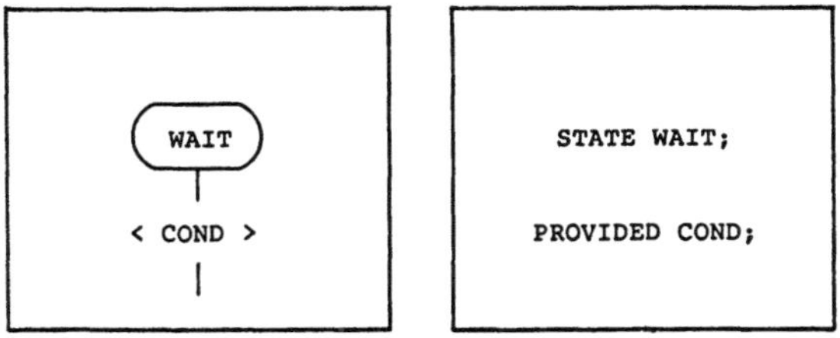

- COND = BOOLEAN EXPRESSION
  IT CAN BE LOCAL OR IMPORTED
- IF COND = TRUE THE TRANSITION IS ACTIVATED
- IF COND = FALSE THE PROCESS INSTANCE WAITS

Fig. 7.6. Continuous signals

Data Concepts in SDL

The data concepts in SDL are based on the principles of abstract data type approach. This means that all data types are defined in an implementation on independent way, only in terms of their properties. The properties are classified in three groups (Fig. 7.7):

— set of values,

— set of operations on these values,

— set of axioms defining the operations.

Each data item has a data type and the data type is defined by the above properties. The set of values is never empty and all values should be enumerated belonging to this data type. The set of operators contains every operator applied to these values. The axioms are the equations defining the behaviour of these operators when applied to the values.

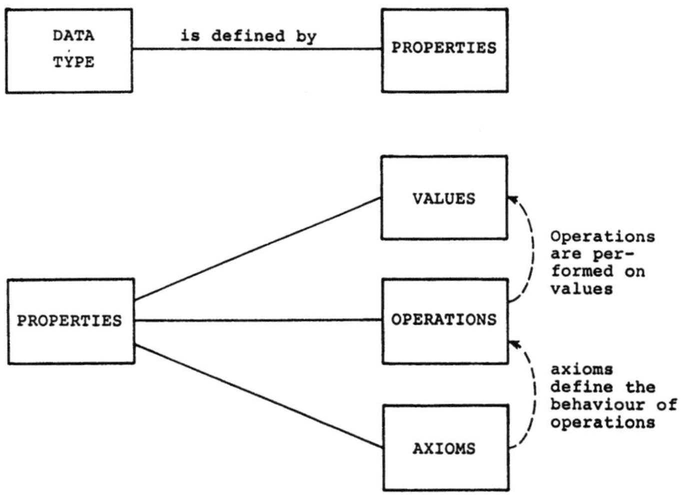

Fig. 7.7. Data concept in SDL

A set of values with certain characteristics is called a sort. Operators are defined from and to values of sorts. For instance the application of the operator for summation ("+") from and to values of Integer sort is valid, whereas summation of the Boolean sort is not. All sorts have at least one value. Each value belongs to one and only one sort, t.e. is sorts never have values in common.

The data types have two big groups:

— predefined data types (Boolean, integer, string, etc.),

— user-defined types.

The predefined types are available to every system. Now we summarize the predefined data types and their properties.

The predefined data types with their set of values and with the set of operations on these values are the following:

— BOOLEAN: usual Boolean data, allowed values TRUE and FALSE. Allowed operations are: AND, OR, NOT.

— INTEGER: usual integer numbers, negative and positive integers. Allowed operations are: "+", "–", "/", "*".
Other operations:
Float: Integer → Real;
Fix: Real → Integer.

— STRING: the data can contain any ASCII character.
The allowed operations are:
Declare!: → String;
"/": String, String → String;
Length: String → Natural;
First: String → Item;
Last: String → Item;
Extract: String, Natural → Item;
Insert: String, Natural, String → String;

— TIME: the data contains a value representing an absolute time expressed as Year, Month, Day, Hour, Minute, Second. The only allowed operation is:
Now → Time.

— DURATION: the data contains a value representing a relative (positive) TIME. Allowed operations are: "+", "–", "*", "/" inherited from REAL, plus the following:
"+": Time, Duration → Time;
"+": Duration, Time → Time;
"–": Time, Duration → Time.

— PID: the data contains a Process Instance Identifier. The allowed operation is: Create!→ Pid (PID values are returned by the interpretation of a create request node, generated by the Create! function).

— CHARACTER: the same meaning as in programming language. The alphabet is CCITT alphabet number 5. The allowed operation is: Ordering!

— POWERSET: it is an ordered set, i.e. the normal mathematical meaning. Allowed operations are:
"In": Item, Powerset → Boolean;
Incl: Item, Powerset → ;
Del: Item, Powerset → ;
Ordering!;

— REAL: usual real numbers. The allowed operations are:
"–", "+", "*", "/", and all defined: Real → Real.

— CHARSTRING: the data contains a string of characters.
  The allowed operations are:
  Declare!→ Charstring;
  "//": Charstring, Charstring → Charstring;
  Length: Charstring → Natural;
  First: Charstring → Item;
  Last: Charstring → Item;
  Extract: Charstring, Natural → Item;
  Insert: Charstring, Natural, Charstring → Charstring.
— TIMER: signal associated at a certain time. Allowed operations are:
  Set: Time → Timer;
  Reset: Time → Timer ;
  Active: Timer → Boolean.
— ARRAY: the data contains an indexed collection of items which are all of the same data type. This is an extension of the normal meaning of array as used in programming language. Allowed operations are:
  Insert!: Array, Index, Item → Array;
  Extract!: Array, Index → Item.
— NATURAL: normal mathematical meaning of natural numbers. Allowed operations are those defined for integers.

The user defined data types are discussed in the CCITT Recommendation Z.104.

Operators

There are two types of operators, the infix operators and the prefix operators. The list of infix operators for predefined operations can be seen in Table 7.3. Only two prefix operators, the "Unary minus" and the "NOT" are used for predefined operations. The user has the possibility to define any prefix operator, as necessary.

Table 7.3. Infix operators

OPERATORS

INFIX	=>	IMPLIES	>=	GREATER OR EQUAL
	OR	OR	+	PLUS
	XOR	EXCLUSIVE OR	-	MINUS
	AND	AND	/	DIVIDE
	IN	SET MEMBERSHIP	*	MULTIPLY
	/=	NOT EQUAL	//	CONCATENATE
	=	EQUAL	MOD	MODULUS
	>	GREATER THAN	REM	REMAINDER
	<	LESS THAN		
	<=	LESS OR EQUAL		

Generally, an operator is a mathematical function mapping one or more values (possibly from different sorts) to a result value. Operators are introduced after the keyword **OPERATORS**. The operators are enumerated and then the signature of the operator follows. The signature contains the sorts of their argument(s) and the sort of result is denoted too.

Examples

a) Operators

```
Take: Primary-colour → Colour;
 Mix: Primary-colour, Colour → Colour.
```

b) Operators

```
   Add: Set, Item → Set;
Delete: Set, Item → Set;
 Is in: Set, Item → Boolean.
```

The operators' section in the data type definition defines the set of values onto which each operator can be applied (the domain) and the set of possible results (the co-domain).

Axioms

The axioms define the behaviour of operations. Axioms are either given as Boolean axioms or in the form of algebraic equivalence equations:

```
         ┌ Boolean expression;
AXIOMS  <
         │ FORALL variable list
         └    IN Type (Axiom list);
```

The axioms follow the signature of the operators as the example shows:

```
Mix(Red, Take (Red)) = = Take(Red);
Mix(Red, Mix(Red, Take (Green))) = =
                         Mix(Red, Take (Green));
```

Another example:

```
FORALL b: bool
    not(not(b))  = b;
    not(true)    = false.
```

Inheritance

The first means of data definition was demonstrated in the earlier part:

a type is defined by giving its properties, that is to say the operations which may be performed on a data item of that type.

The second way is to define a type by referring to a set (or subset) of operations defined in conjunction with another type. The type being defined in this fashion inherits some or all of the properties of the type referred to in the definition (known as parent type).

The parent type can have defined through a **GENERATOR**, a **NEWTYPE** or it can be a predefined type. The following scheme shows the possible relations:

PARENT TYPE		DERIVED TYPE
GENERATOR	⟶	NEWTYPE
	⟶	GENERATOR
NEWTYPE	⟶	NEWTYPE
	⟶	SYNTYPE
PREDEFINED	⟶	NEWTYPE
	⟶	SYNTYPE
SYNTYPE	⟶	SYNTYPE

The **PREDEFINED** type was earlier discussed, the others have the following definition:

— The **GENERATOR** is a valuable construct allowing the definition of an abstract type that can be instantiated in different ways.

— The **NEWTYPE** construct allows the definition of a type. The type is defined by giving the values for that type (**CONSTANT** construct) and the set of operators (properties) that can be used to manipulate that type.

— The **SYNTYPE** construct defines a type having a subset of the values of the parent data type.

The use of the **GENERATOR** is explained by the following example:

```
ARRAY DATA TYPE

     GENERATOR   ARRAY ( TYPE INDEX? TYPE ITEM);

       OPERATORS

            <ASSIGNMENT>;

            DECLARE ! : —→ARRAY;

            INSERT !  : ARRAY' x INDEX x ITEM —→ ARRAY;

            EXTRACT ! : ARRAY x INDEX —→ ITEM;

       AXIOMS

            EXTRACT ! (DECLARE, I) = ERROR !;

            EXTRACT ! (INSERT! (A, IPos, It), EPos) = If EPos =

                         IPos THEN It ELSE EXTRACT! (A, EPos) FI;

   END      ARRAY
```

This example shows a possible definition of a type generator "Array".

The type generator "array" leaves undefined the two types composing it.

The role of these two types (where "index" represents the maximum length of the array and "item" represents the type of its components) may be understood by looking at the operators and axioms parts.

The generator aims at providing a model valid for a variety of types referencing to that model. In our example it will be possible to define arrays having different length (from 1 up to 20 or from A to Z) and containing different types of elements (Integers, Characters etc.).

The operations allowed are Assignment, Declare, Insert and Extract.

The operations are such that Declare produces:

— an array,
— the Insert, given the array,
— the position in the array and
— the item produces a new array and
— the extract, given an array and a position, produces an item.

These operations are defined by the following axioms where:

Extract! (Declare!,I) = Error!

indicates that it is an error attempting an extraction from an empty array (an array just create where no insertion has been made).

```
Extract! (Insert! (A,Ipos,It), Epos) =
     IF Epos = Ipos THEN It ELSE Extract!(A,Epos)
```

indicates that the extraction from an array A of an item in position Epos, gives as result the item It having been inserted in that position (Ipos=Epos).

The use of **NEWTYPE** is demonstrated by an example, but initially we should mention the following principles:

It is possible to define a **NEWTYPE** inheriting properties from an already defined **NEWTYPE**. It is not allowed to inherit properties from a **SYNTYPE**.

The inherited properties are listed by name after the keyword **INHERITS**. If the **NEWTYPE** has to inherit all the properties the keyword **ALL** is used.

It is also possible to add new properties to the set of those inherited: in this case they are listed after the keyword:

ADDING OPERATORS.

The new properties have to be defined to complete the type definition.

A **NEWTYPE** may be derived from existing types, in our example we derive the POUND(£) type from **INTEGER**.

```
NEWTYPE POUND
     INHERITS INTEGER ("+","-")
     ADDING OPERATORS
        "*" = POUND x INTEGER ——→ POUND
END POUND
EXAMPLES
                  £ 3 + £ 2               = £ 5
     BUT          £ 3 +   2 (INTEGER)     = NONSENSE
                  £ 3 * £ 2               = NONSENSE
     HOWEVER      £ 3 *   2               = £ 6
```

The values of **POUND** are distinguished from those of the **INTEGERs** by prefixing a £ sign.

The **POUND** type inherits addition and subtraction from integers, but not division or multiplication.

Although it is possible to multiply pounds by integers, it is not allowed to add an integer to a pound nor to multiply pounds by pounds.

As is clear from this example the inheritance of properties allows much more compact and easier definitions: in fact if we choose to state all the properties of the POUND type without referring to the INTEGER type we would have to state all the axioms related to the operators "+" and "–".

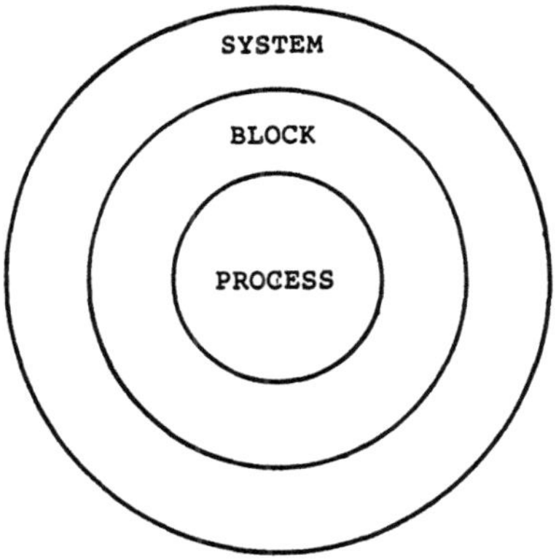

Fig. 7.8. Data hierarchy

Finally the data hierarchy is demonstrated in Fig. 7.8. The data of a process can be exported or revealed. The general rules for the hierarchy of data visibility boundaries are the following:

— predefined types are available to every system,
— system defined types are available to all blocks of the system,
— block defined types are available to all processes of the block,
— process defined types are only available to the defining process,
— process defined data are local to that process unless explicitly exported or revealed.

An item of data is manipulable by the owner:

— visible by viewers of the block if revealed,
— importable by anyone if exported.

There are two pairs of attributes for this purpose:

REVEAL/VIEW and EXPORT/IMPORT

The REVEALING attribute implies that the value of that data object is also visible in the block containing that process, and therefore visible to all processes of the block provided they declare the same data as having the VIEWING attribute.

A data object declared as EXPORTED in a process can be IMPORTED by processes within the same block and outside the block.

Tools for SDL

In principle, SDL can be used without any tools. However, the inherent complexity of modern systems is such that the SDL specifications are often complex.

Therefore, automatic tools are needed to help manage the specification, design, and documentation of the systems. The application field of the SDL tools concern the system modelling and the document handling.

In this section we shortly outline a set of tools which may support SDL (Fig.7.9).

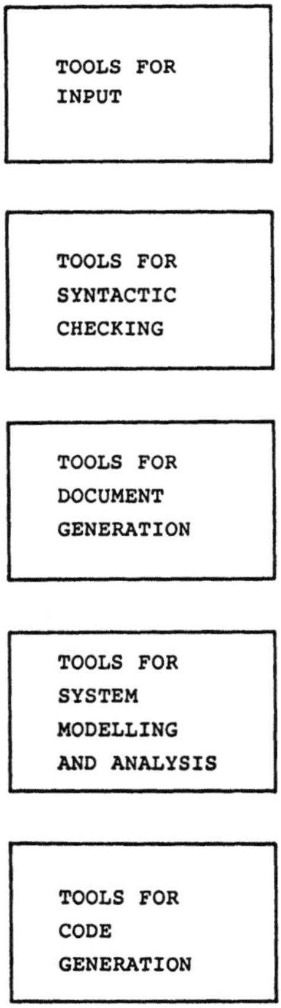

Fig. 7.9. Tools for SDL

Tools for input: according to the syntactic forms we have input aids for the SDL graphic, textual phrase, and pictorial, No special requirements are needed for entering the textual phrase form of SDL, as the SDL/PR is equivalent — from the input viewpoint — to any character string input. The other syntax, however, assumes graphic a predefined grid and associate to each symbol a given string. Some graphics devices and a graphics editor are useful as a means of outputting diagrams in graphic form.

Tools for syntactic checking: this class includes syntax analysers, valid for each of the syntaxes.

Tools for document generation: Once SDL documents are stored within a machine, tools can access and reproduce them, possibly using several peripherals. They may use a syntactic form different from that used to input the document. In addition,tools may be able to generate new types of documents derived from those originally entered.

Tools for system modelling and analysis: The SDL documents representing a system can be used for abstracting a model of the system. Checks can be performed on that model. They can search for deadlocks, make a comparison between various models of the same system, or run a simulation of the system behaviour.

Tools for code generation: The SDL representation can automatically or semiautomatically (interaction with a designer) be inspected and a programming code performing the required behaviour can be generated from SDL.

Finally, some remarks are made relating to the document generation.

Starting from the SDL documents input by users and stored in the machine, which we call "primary", several other documents can be generated automatically, among these:

- The *signal lists* organized per processes, per block, per system.
- The *process interaction diagrams* showing the interactions and the following sequences of actions in communicating processes.
- The *state overview diagrams* showing the process graph as a set of states connected by arcs representing the transition.
- The *cross reference* tables organized per process,per block,per system.
- The *partitioning diagram* showing the blocks and levels structure.
- The *sequence charts* showing a system (process) behaviour as responses to certain sequences of environmental actions.
- Indexes and Dictionaries or Glossary.

The documentation structure is demonstrated in Tables from 7.4 to 7.7: Table 7.4 shows the basic system documentation list, Table 7.5 — the block documentation, Table 7.6 — the process documentation, Table 7.7 — the procedure documentation.

Table 7.4. System documentation structure

SYSTEM		
	DATA TYPE DEFINITION SYNONYM DEFINITION PROCEDURE DEFINITION	TYPES, SYNONYMS, PROCEDURES VALID FOR ALL THE SYSTEM
	SIGNAL DEFINITION	SIGNALS INTERCHANGED WITH THE ENV AND BETWEEN SYSTEMS BLOCKS
	CHANNEL DEFINITION	CHANNELS CONNECTING SYSTEM'S BLOCKS TO EACH OTHER AND TO THE ENVIRONMENT
	BLOCK DEFINITION	DEFINITION OF SYSTEM'S BLOCKS
ENDSYSTEM		

Table 7.5. Block documentation structure

BLOCK		
	DATA TYPE DEFINITION SYNONYM DEFINITION PROCEDURE DEFINITION	TYPES, SYNONYMS, PROCEDURES VALID FOR THE BLOCK
	SIGNAL DEFINITION	SIGNALS INTERCHANGED WITHIN THE BLOCK
	PROCESS DEFINITION	PROCESSES OF THE BLOCK
	STRUCTURE DEFINITION	BLOCK DECOMPOSITION
ENDBLOCK		

Table. 7.6. Process documentation structure

PROCESS	NAME, INSTANCES	
	DATA DEFINITION	VARIABLES, SYNONYMS AND DATA TYPES LOCAL TO THE PROCESS
	PROCEDURE DEFINITION	PROCEDURES LOCAL TO THE PROCESS
	PROCESS DIAGRAM	
ENDPROCESS		

Table 7.7. Procedure documentation structure

PROCEDURE		
	SIGNAL VISIBILITY	
	DATA VISIBILITY	
	DATA DEFINITION	VARIABLES, SYNONYMS AND DATA TYPES LOCAL TO THE PROCEDURE
	PROCEDURE DEFINITION	PROCEDURES LOCAL TO THE PROCEDURE
	PROCEDURE DIAGRAM	
END PROCEDURE		

7.2 ARCHITECTURAL SPECIFICATION CONCEPT

The development of the description languages started nearly parallel with the Reference Model development in ISO. The goal was to get an unambiguous, concise and clear interpretation for protocol and service specification.

Earlier the protocol specification was based on two different models. The control structure of protocols was specified using the state machine approach. The semantic content was described by the programming language approach. This approach was used with properties involving counting and, in general, in cases when the state machine representation would become too complex. The basic idea was to create a unified model [BOCH-GE-77] from these two methods in order to get a universal tool for protocol specification and verification.

Two description languages have been developed in ISO, both languages have a common architectural specification concept [VI-TE-BO-83]. Protocols, services, interfaces and the entire OSIA are concerned as systems. The architecture of a system (Fig. 7.10) is composed of a set of interacting modules and the modules are interconnected with channels. Each channel has two endpoints, called interaction points. The basic notions of the architectural specification concepts are the module, the channel and the interaction point (Fig. 7.11).

The *modules* embody the actions. The actions are exclusively allocated to modules. A module is a unit of specification. The behaviour of the module can be observed at the interaction points. One description language of ISO, called ESTELLE, is based on the extended finite-state machine concepts. It means that every module is specified as an extended finite-state transition.

Modules interact with one another through *channels*. A channel is realized by an "abstract interface". A module has contact with its environment only through a well-defined set of channels. A channel connecting two modules could be specified by naming the interaction points (e.g. T-SAP, S-SAP).

Each *interaction point* of a module is characterized by the type of channel which it represents. A channel type can be specified enumerating the possible interaction primitives, naming the two channel ends and the connected modules and giving the sequence order of the execution. The other description language of ISO, called LOTOS, is based on the temporal ordering of interaction primitives.

The common architectural specification concept of ESTELLE and LOTOS helps the interworking of the two languages.

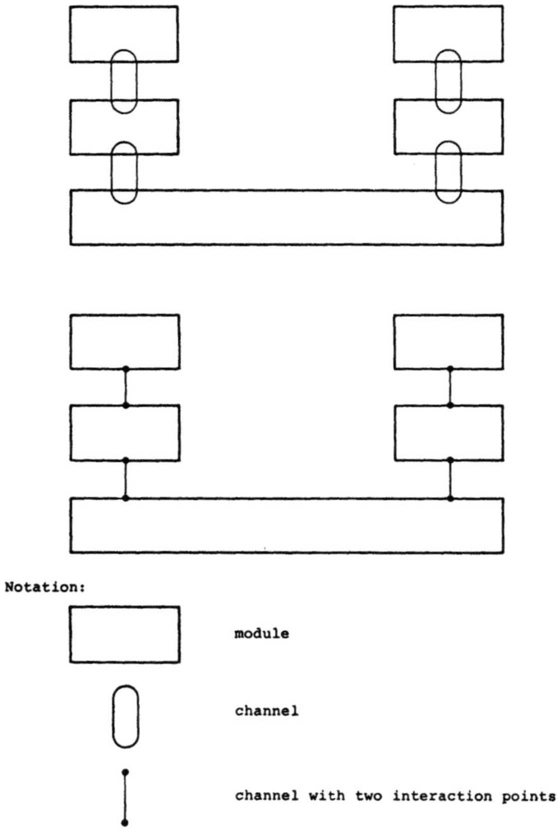

Fig. 7.10. Architecture of a system

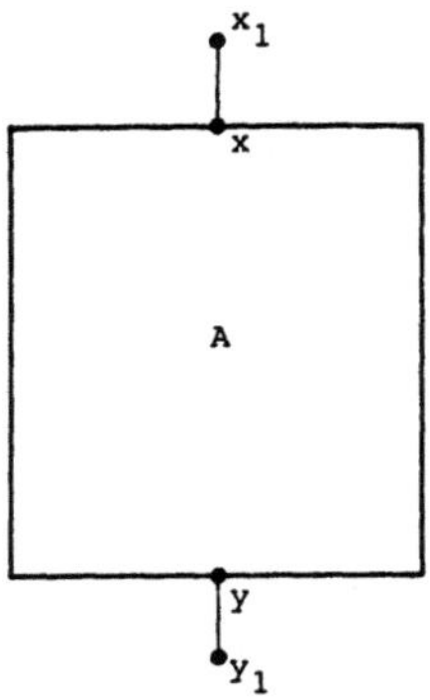

Module A

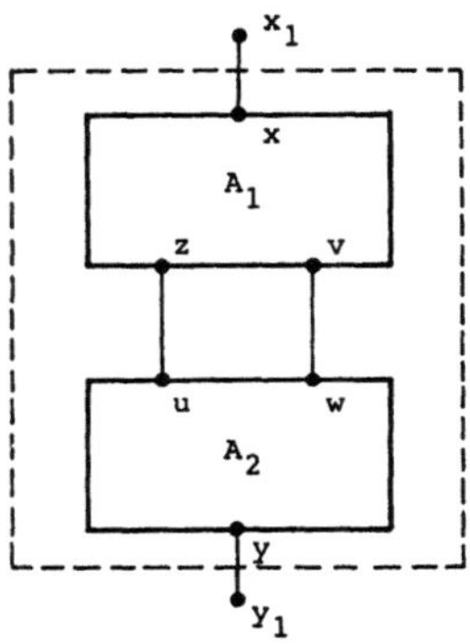

Modules A_1 and A_2 representing the substructure of A

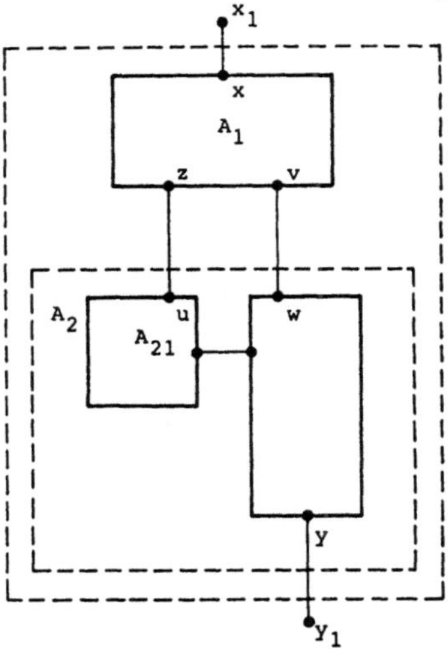

Module A_1 and module A_2 with submodules

Fig. 7.11. System modules

7.3. ESTELLE, AN EXTENDED FINITE-STATE MACHINE SPECIFICATION LANGUAGE

Overview of ESTELLE

ESTELLE is a formal language for specifying distributed systems with a particular application of protocols and services. ESTELLE can be used for abstract specification as well as for the (implementation, testing, etc.), of real-world tasks.

ESTELLE contains two components, the model and the language. The model is based upon an extended finite-state machine model. The extended approach means that the simple finite-state machine description is combined with the power of a programming language. The language is largely based on the syntax and semantics of the PASCAL programming language.

The ESTELLE model has four main levels of abstraction (Fig. 7.12). The highest level is the specification. The next level of the hierarchy is composed of modules. The modules can be divided into transitions. A transition contains the actions and processes.

Essentially this hierarchy creates the first principles of the ESTELLE language (Fig. 7.13).

The main features of ESTELLE [LINN-86, AN-CH-DEM-86, COU-PE-AY-85] will be introduced from transitions up to the specification [COU-PE-AY-85].

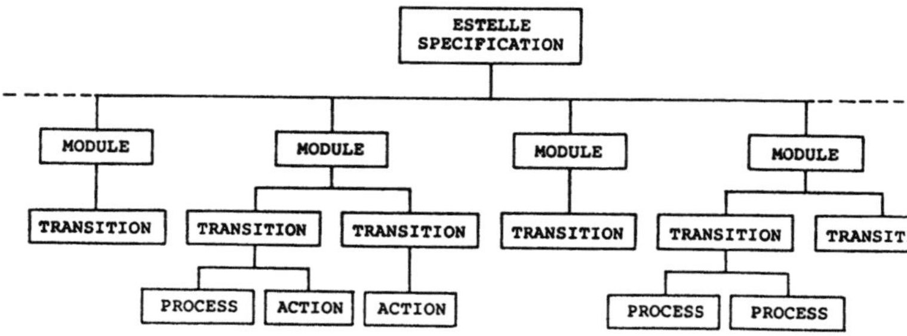

Fig. 7.12. Hierarchy of the ESTELLE model

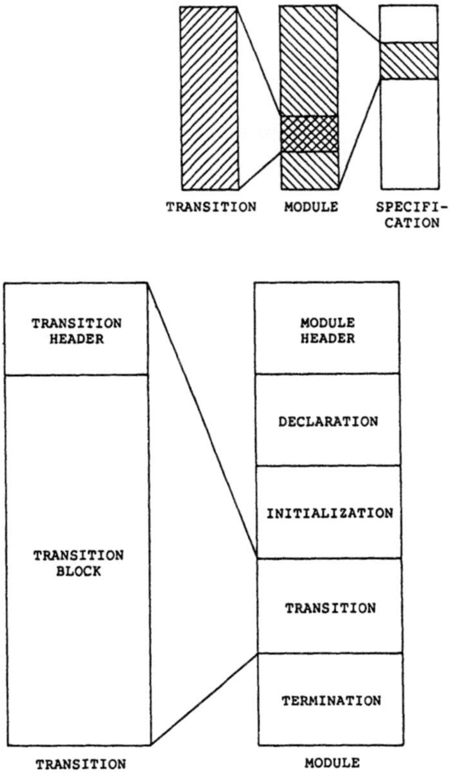

Fig. 7.13. Internal structure of the ESTELLE model

Transitions

A finite-state transition table is very often the starting point of a protocol specification. The table identifies the state names, state transitions, criteria for a transition from one state to another and identifies actions to be taken when a transition fires.

The general form of an ESTELLE transition definition is the following (the keywords are underlined)

```
trans
  priority    expression (for the transition
                          in relationship to
                          other transitions)
  from        state-a    (current state)
  to          state-b    (next state)
  provided    predicate  (boolean expression)
  when        inp.event  (input required)
     begin
              (transition block)
     end
```

A transition specification has two parts, the transition header and the transition block. The transition specification begins with the keyword *trans*. Every factor to be considered when a transition from one state to another occurs is expressed in the transition header. If all of the conditions expressed in the clauses of a header are true, then the actions specified in a transition block are executed and the next state is selected.

The main conditions are the priority, the predicates and the required input events. The *priority* represents the first decision to be evaluated. If we have a set of transitions, the priority-expression determines the sequence of transitions. The predicates are described following the keyword *providing*. The number of enabling predicates is unlimited. The *when* clause introduces an input interaction. All three conditions are optional. The absence of the *when* clause represents a spontaneous transition. A *delay* value often follows the *when* keyword giving the minimal and/or maximal delay of the spontaneous transition.

The essential part of the transition header is determined with the *from* and *to* keywords.

```
trans

     from WFTRESP
          to CLOSED
               when TC_IP.NDIS_IND
                    name TC15:
                    begin
                    output TCEP.TDIS_IND;
                    TC_RELEASE_REQ:=TRUE
                    end;

          to WFND
               when TCEP.TDIS_REQ
                    name TC14:
                    begin
                    BUILD_TCC(TC_DREF,PDU_SEND);
                    output TC_IP.TPDU(PDU_SEND)
                    end;

          to OPEN
               when TCEP.TCON_RESP
                    name TC13:
                    begin
                    BUILD_TCA(TC_DREF,
                              TC_SREF,
                              TC_CALLING,
                              TC_CALLED,
                              TC_MXBL_S,
                              PDU_SEND);
                    output TC_IP.TPDU(PDU_SEND)
                    end;

     from WFTRESP
          to ERROR
               when TC_IP.INVALIDPDU
                    name TC21:
                    begin
                    ITPDU:=INVALID;
                    output TCEP.TDIS_IND
                    end;

     from WFTRESP
          when TC_IP.TPDU
               provided PDU.CODE<>TBR
                    to ERROR
                         name TC16_17_18_19:
                         begin
                         output TCEP.TDIS_IND;
                         D_NULL(ITPDU)
                         end;

               provided PDU.CODE=TBR
                    to CLOSED
                         name TC20:
                         begin
                         output TCEP.TDIS_IND;
```

Fig. 7.14. ESTELLE description of a small transport protocol fragment

The *from* clause introduces the name of the state to which the transition belongs; i.e. the machine must be in this state for the transition to be selected and is called the "current state" of the machine. The *to* clause identifies the state of the finite-state machine after selection and execution of a transition. This is the "next state".

The second functional part of the transition is the transition block enclosed between *begin* and *end* keywords. The content of this block is the action. The action can be performed at transition firing, generating one or several output interactions.

A small fragment of the Transport protocol Class 2 and its ESTELLE description can be seen in Fig. 7.14.

Modules

The module is the next higher level of abstraction. The module is the central notion of ESTELLE. Modules communicate by exchanging messages (called: interactions) through bidirectional channels connecting their ports (called: interaction points). A module is a black box containing the transitions and ensuring the connection with the environment through its interaction points, exported variables and parameters.

A module has a header and a body. The header defines the header-type, the list of interaction points, parameters and exported variables. The body specifies the internal behaviour of the module as well as its submodules.

The general form of an ESTELLE module header definition is the following.

```
module header-type- identifier module-class
       (interaction-point-list;
       param parameter list;
            export exported-variable-
                   list; )
```

The header begins always with the keyword *module*. The header-type-identifier helps us to give the module a mnemonic name (e.g. header-type is "Transport"). The module-class is either process or activity. Processes may run in parallel with other processes at the same level of the hierarchy. Activities must be at the lowest level of a subtree in the specification hierarchy.

The next line of module header definition is the list of interaction points. The interaction points are the abstract interfaces between the modules. Each interaction point should have a name in the module header (e.g. TSAP]. Interaction points are full duplex interfaces: they allow independent input and output interactions to be exchanged asynchronously with other modules. The interactions are the data to be exchanged (e.g. service primitives). The channel type definition belongs to the definition of interactions.

The parameter list starts with the keyword *param*. Both constants and variables can be specified in the parameter list with their associated data type. Exported variables are prefixed by the keyword *export* in a module header. Exported variables are shared between parent and child modules or shared by sibling modules which are activities.

The module body consist of the following parts
— a declaration part,
— an initialization part,
— a transition-declaration part,
— a termination part.

Each part is optional. The module body starts with the keyword *body* preceding the name of the module body. The declaration part includes the definition of submodules and specifies constants, types, channels, control states and variables. The declaration is very similar to Pascal.

The initialization part is prefixed with the keyword *initialize*. The initialization part may establish initial values to declared objects and may create some submodules and channel links between them.

The transition-declaration part was discussed in the previous pages.

The termination part defines a procedure which is automatically executed before a module and its resources are released.

Figure 7.15 shows an example of a module header and body [AHO-SAR-87b].

```
(* Module Header *)

MODULE  H_Type_Id  "class"  ("Interaction_point_decl"
                              "Parameter_List");
                              "Exported_Vars"


(* Module Body *)

BODY  B_Id  FOR  H_Type_Id;

  "Declaration Part"
      (* Const_Declaration_Part
         Type_Declaration_Part
         Channel_Type_Definition
         Module_Type_Definition
         Variable_Declaration_Part
         State_Set_Definition_Part
         Use_Clause
         Procedure_And_Function_Definition_Part *)

  "Initialization_Part"

  "Transition_Declaration_Part"

  "Termination_Part"
```

Fig. 7.15. Module header and body

Interaction Points and Channel Type

The ISO architectural specification concept is based on modules, channels and interaction points. The two ISO-languages, ESTELLE and LOTOS, differ from each other in the interpretation of modules, channels and interaction points. We have seen ESTELLE's module definition in the previous pages, we now deal with the rest.

An *interaction point* is an abstract interface of a module used to exchange interactions with other modules. There are external and internal interaction points. The *external* interaction points are specified in the module header definition to exchange interactions with its external environment. For instance User and Provider are typical interaction points and they are specified in the module body definition to exchange interactions with the internal environment of the module. A typical *internal* interaction point is a timer interface.

Interaction primitives may be messages consisting of structured data. Protocol data units exchanged between adjacent layers via service primitives.

Each interaction point has an associated FIFO queue to which input messages are appended. The queueing discipline determines if it is an individual or a common queue. The individual queue belongs to an individual interaction point, the common queue is shared among more interaction points.

The *channel* definition contains the channel type identifier, the role list and the objects to be exchanged. The channel type identifier describes the type of interaction point (e.g. User-access-point). The role list defines generally the appropriate service access point with two roles, User and Provider. The objects are the primitives with their parameters and abstract data type (e.q. T-CONNECT-rec, TCEP-identifier, to-T-address, from-T-address, QOTS-request, TS-connect-data).

Specification

The ESTELLE specification is the highest level, similarly to a program in Pascal. An ESTELLE specification defines a system of hierarchically structured and communicating modules or module instances.

The typical structure of an ESTELLE specification is demonstrated in Fig. 7.16. The entire specification is embraced between the keywords *specification* and *end*. The first, declaration part is mandatory and it contains the global definition of the environment. The second initialization part is optional and it is responsible for the initialization of child processes.

The next big part is the specification body composed either of a complete self contained module or of a set of nested modules. The nesting of modules implicitly identifies one relationship between the modules. A module defined within the scope of another module is only visible within the scope of the embodying module. The embodying module is called parent, the embedded nested module is called child module. A specification is partitioned into a set of modules. The modules can be nested within each other forming a tree of modules organized in a hierarchy. Two examples show the ESTELLE specification going into detail. The first example is in the APPLICATION part in Section 8.5. It specifies the alternating bit protocol.

```
body SENDER for SENDER_type;

var
 saved_data    : marked_data;
 Idle,Wait     : boolean;

function encode_DT(m_data: marked_data): data; external;

procedure decode_AK(var ak:logical_ak_type;n_data:data);external;

initialize
  begin
   saved_data.seq := 0;
   Idle := true;
   Wait := false;
  end;

trans when U.Transmit_Request
     provided Idle
        begin
         keep_copy (U_data, saved_data.u_data);
         output N.NSDU_Request ( encode_DT (saved_data) );
        end;

trans when N.NSDU_Indication
        begin
         decode_AK (received_ak,N_data);
         if Wait and (received_ak.present) and
             (received_ak.seq =saved.data.seq)
           then
            begin
           output U.Transmit_Confirmation;
           saved_data.seq := ( saved_data.seq +1 ) mod 2;
           received_ak.present := false;
           Idle := true;
           Wait := false;
            end;
        end;
trans provided Wait
        delay (Retran_time)
        begin
         output N_.NSDU_Request (encode_DT(saved_data));
        end;

end; (*end of SENDER body*)
```

Fig. 7.16. Typical structure of an ESTELLE specification

Table 7.8. An Example (a part of a module without substructure)

```
MODULE Recl_Type PROCESS (List2 .... List3 .... List4 ....;
                          List6: List6_id(Receiver) INDIVIDUAL QUEUE;
                          PARAM Port_Nr.: Port_Type);

BODY  Recl_Body  FOR  Recl_Type;
  ....
  VAR  V_S,V_R: Seg_Type;
  ....
  TRANS
    FROM  Transfer  TO  SAME
      WHEN List6.Dat
        PROVIDED R_Nr ≠ V_R + 1
          BEGIN  OUTPUT List3.Rej(V_R);  END
        PROVIDED R_Nr = V_R + 1
          BEGIN  V_R:= V_R + 1;
                 OUTPUT List3.Rr(V_R + 1, S_Nr);
                 OUTPUT List2.U_Dat(I);
          END

    FROM  Transfer  TO  Discon
      WHEN List6.Dis
           BEGIN  END
  ....
```

Example

A short example [AN-CHA-DEM-86] demonstrates the usage of ESTELLE. The example describes a simple protocol through which a number of users communicate with a network. We assume for simplicity that each connection has its separate process both for sending and receiving functions. The internal structure consists of the user processes senders and receivers running parallel (Fig. 7.17). This architecture is defined by the initialization part of the Entity module. In this part instances are created and their interaction points U (User) and N (Network) are attached to corresponding Entity's interaction points.

The Entity itself is not an active module (no transitions) and it serves only as a logical interface unit between users and network. Its internal structure could have been different. The external context in which the Entity is defined is given by the constant, type and channel definitions preceding the Entity specification.

Figure 7.18 shows the operation of the sender protocol. After a user request has been received (Transmit-Request interaction through the interaction point U) the PDU DT with the subsequent message number is sent(output NSDU-Request through N) and the sender waits for acknowledgement. If a proper acknowledgement arrives in time then the message number is incremented and a confirmation sent (Transmit-Confirmation) restarting the whole process again. If a certain time has passed without acknowledgement the saved (saved-data variable) user PDU is retransmitted.

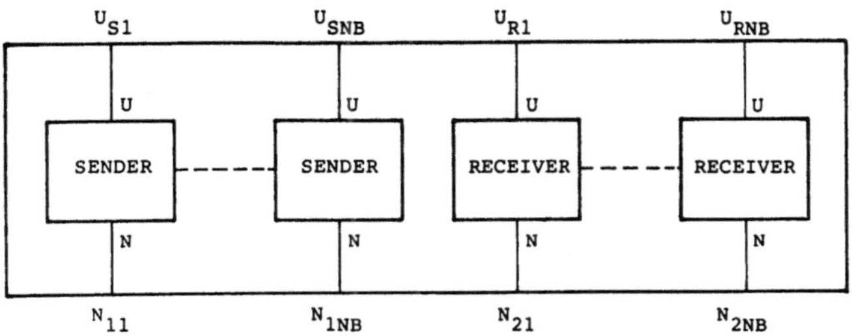

Fig. 7.17. Entity

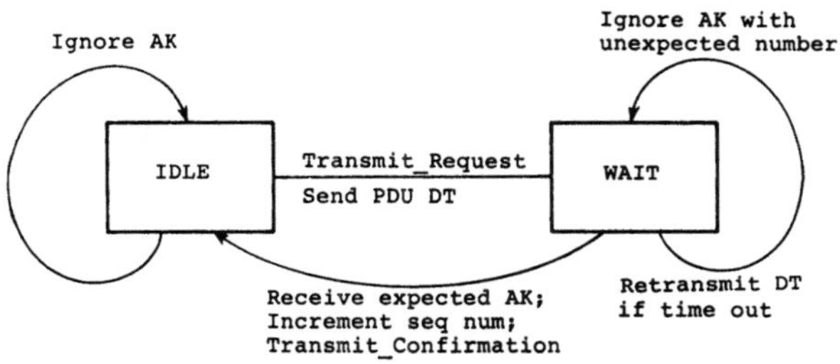

Fig. 7.18. Sender protocol

This short English description is very naturally translated into ESTELLE transitions with two control states, IDLE and WAIT, separating the wait for the user data and network acknowledgement, respectively.

The ESTELLE specification of this simple protocol can be seen in Table 7.9.

Table 7.9. ESTELLE specification of a simple protocol

```
default individual queue;

const
 Retran_time = any integer;       (*retransmission timer value*)
 NB          =  any   integer;        (*number   of   protocol
machines*)

type
 Seq_type = 0..1;                 (*sequence number*)
 data     = ...;                  (* data *)
 logical_ak_type =
          record
            present : boolean;    (*an ak has been received*)
            seq     : seq_type;   (*the sequence number*)
          end;
 marked_data = record
                u_data : data;     (* user data received *)
                seq    : Seq_type; (* sequence number *)
               end;

channel U_SENDER_channel (User, Provider);   (*sender channel*)
 by User:
  Transmit_Request ( U_data : data);
 by Provider:
  Transmit_Confirmation;

channel N_channel (User , Provider);         (*network channel*)
 by User:
  NSDU_Request ( N_data : data );
 by Provider:
  NSDU_Indication ( N_data : data );

channel U_RECEIVER_channel (User, Provider); (*receiver channel*)
 by User:
  Receive_confirmation ;
 by Provider:
  Receive_Indication ( U_data : data);

procedure keep_copy ( to_be_saved: data; var  copy:data);  primi-
tive; (*keep a copy of some data*)

module Entity_type;
 ip  US :array [1..NB] of U_SENDER_channel (Provider);
     UR :array [1..NB] of U_RECEIVER_channel (Provider);
     N  :array [1..2,1..NB] of N_channel (User); end;

body   Entity for Entity_type;

   module RECEIVER_type process;
    ip  U: U_RECEIVER_channel(Provider);
        N: N_channel(User);
   end;
   body RECEIVER for RECEIVER_type; external;

   module SENDER_type process ;
    ip  U: U_SENDER_channel(Provider);
        N: N_channel(User);
   end;
```

Table 7.9 (continued)

```
body SENDER for SENDER_type;

var
 saved_data    : marked_data;

state
 Idle,Wait;

function encode_DT ( m_data: marked_data) : data; external;
 (* encode a DT pdu containing
      - a PDU code ( identifying a pdu of category DT)
      - the sequence number
      - the user data *)

function decode_AK (n_data : data) :logical_ak_type;external;
 (* decode an NSDU and recognize an AK; present field
    true if a correct ak is received, false otherwise *)

initialize (*of sender body*)
 to Idle
  begin
   saved_data.seq := 0; (*initial seq number*)
  end;

(* the User requests the transmission of a message *)
trans  when U.Transmit_Request
         from Idle to Wait   (*request is valid*)
          begin
           (* the data passed by user are locally saved
              for later retransmission if necessary*)
           keep_copy (U_data, saved_data.u_data);
           (* encode_data returns an encoded PDU *)
           output N.NSDU_Request ( encode_DT (saved_data) );
          end;
(* NSDU received from network *)
trans when N.NSDU_Indication
        from Wait
          provided (decode_AK (N_data).present) and
                   (decode_AK (N_data).seq = saved_data.seq)
            to Idle (*an expected_ak has been received*)
              begin
                output U.Transmit_Confirmation;
                saved_data.seq := ( saved_data.seq +1 ) mod 2;
              end; (*ak has been processed*)
          provided otherwise
              begin
              end; (*ignore wrong AK*)
        from Idle
              begin
              end; (*ignore unexpected nsdu*)
(* retransmission timer *)
trans from Wait
       delay (Retran_time)
          begin
           output N.NSDU_Request (encode_DT(saved_data));
          end;

end; (*end of SENDER body*)
```

Table 7.9. (continued)

```
.bp
modvar
 S: array [1..NB] of SENDER_type;
 R: array [1..NB] of RECEIVER_type;

initialize (*of Entity*)
 begin
 all X: 1..NB do (*initialize all senders*)
  begin
   init S[X] with SENDER;
   attach US[X] to S[X].U;
   attach N[1,X] to S[X].N;
  end;
 all X: 1..NB do (*initialize all receivers*)
  begin
   init R[X] with RECEIVER;
   attach UR[X] to R[X].U;
   attach N[2,X] to R[X].N;
  end;
 end;
end; (*end of Entity body*)
```

Software Tools for ESTELLE

ESTELLE is easy to use *with* some software tools. ADI and Bull in France, and the National Bureau of Standards and the Protocol Development Corporation in the USA, have developed commercial ESTELLE tools.

The software tools for ESTELLE [AN-CHA-DEM-86] can be classified into three groups:

— programming aid tools [AN-87]

- editor
  it creates an intelligent programming environment to help build and modify an ESTELLE specification
- compiler
  it is the tool for parsing a specification, for performing all the static checking and for producing an output representation of an ESTELLE specification
- debugger
  it helps to interpret an ESTELLE description under the user control

— verification tools [SID-BLU-86]

- finite-state machine models
- Petri Net based models

Both verification tools can generate FSM tables for a protocol entity and verify certain protocol properties

— simulators [COU-PE-AY-85]

they are complex tools for experiments, tests and verification. Generally the simulator environment is composed of the simulated model and the user interface module.

7.4 LOTOS, A LANGUAGE OF TEMPORAL ORDERING SPECIFICATION

Overview of LOTOS

Formal description techniques are methods of defining the behaviour of a system in a language with a formal syntax and semantics, instead of a natural language such as English. The ISO defined two languages, the ESTELLE (see Section 7.3) and LOTOS for this purpose.

LOTOS (Language of Temporal Ordering Specification) was designed for application in the protocol field [ISO IS 8807]. The basic idea that LOTOS developed from was that systems can be described by defining the temporal relation between events in the externally observable behaviour of a system. The formal mathematical model of LOTOS has two components:

— the Calculus of Communicating Systems (see Section 6.1),
— the abstract data type language ACT ONE (see Section 6.2).

The CCS deals with the description of process behaviours and interactions. Its basic elements are the process and the event. A process is a system component that may interact with its environment (e.g. other processes). The event is the atomic form of interaction between two processes. LOTOS uses a modification of CCS, called CCS*. It is a finite variant of CCS.

ACT ONE deals with the description of data structures and value expressions. It describes the abstract relationships among data structures and not the concrete representations. This abstract syntax may be realized by different concrete syntaxes [GOTZ-87].

This section will explain the interactions between processes, how processes can offer events to each other and a simple example will show a process specification. Then the syntax and semantics of LOTOS will be summarized and every operator will be briefly discussed. Finally, some examples demonstrate the power of LOTOS [BOL-DENI-86], [BRI-86], [BRI-FE-LO-87].

This section assumes a knowledge of the basic ideas of CCS and ACT ONE (see Sections 6.1 and 6.2).

Interactions Between Processes

The basic idea of LOTOS is to describe systems in terms of their capability to interact with their environment. The system is a black box and its behaviour can be observed only through the points of interactions during a communication event (Fig. 7.19).

The system is described in terms of processes. The system as a whole is a single process that may consist of several interacting subprocesses. These subprocesses may be decomposed into sub-subprocesses, etc. Essentially, the specification of a system in LOTOS is a hierarchy of process definitions.

The specification of system behaviour is based on interaction between processes. The atomic form of an interaction is described as an event. The points of interaction are called event gates in LOTOS (Fig. 7.20).

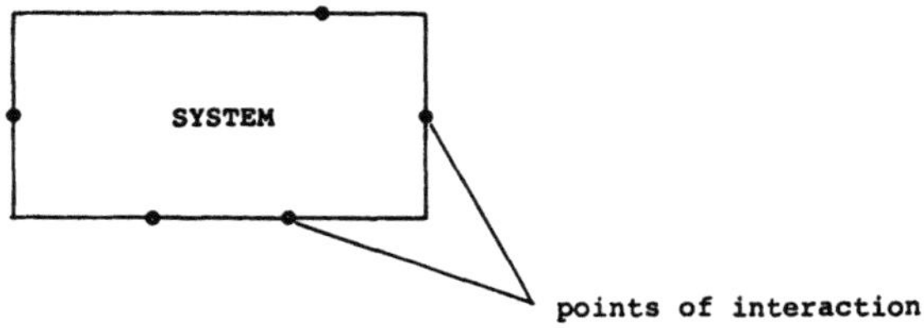

Fig. 7.19. A system in LOTOS

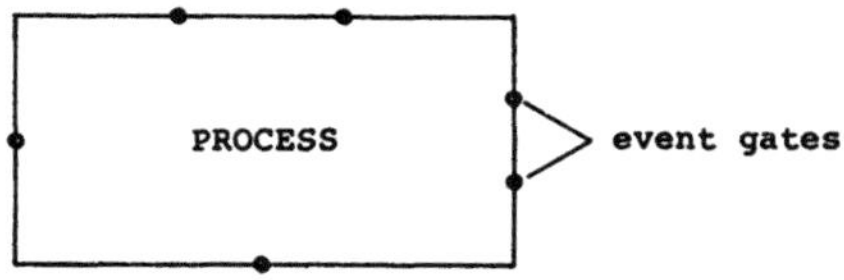

Fig. 7.20. A process in LOTOS

The notation of an event offer is the following:

g ! E

where
- g is the gate name,
- E is an expression,
- ! is an output operator denoting that the gate wants to send a message.

The interaction between processes has special rules. An event may occur if two event offers match
— the events are offered at the same time,
— the events are offered at the same gate,
— the expressions have the same value.

If process A and B have matching event offers, then the two processes will be synchronized. An input event can also be assigned to a gate, its notation is

g ? E

where
- ? is an input operator denoting that the gate is open to receive a message

If all values of a type "t" are offered at the same time, this may be abbreviated as

g ? x : t

where

x is a variable
t is the sort of x

Example:

t = Boolean
"g ! true [ ] g ! false" may also be
written as "g ! x : Boolean"
Where [ ] is the choice operator (explained in the next section)

The event is the basic unit of the specification of interactions. An event is an atomic action, it cannot be interrupted. The input and output events are the most simple observable events. LOTOS has an internal action, denoted by i. This internal event can specify actions inside the process.

If an event occurs, it can be interpreted as a value passing, e.g

process A offers g_1 ! 10
process B offers g_2 ? x : integer

The offer of process A matches with one offer of process B, the choice is resolved for x = 10.

The different kinds of interactions between processes are listed in Table 7.10.

The central notion of LOTOS is the process. The format of a process definition is:

```
process     < process-indentifier >
            < parameter-part >: =
            < behaviour-expression >
endprocess
```

The process-identifier is the name by which the process can be referenced. The parameter-part enumerates the gates and parameters of the process and qualifies the type of variables. The behaviour expression defines the observable behaviour of the process.

A small example from [BOL-DENI-86] demonstrates the process definition. The process has three gates. It accepts two natural numbers at two input gates, in any temporal order, and then offers the larger of them at an output gate.

Table 7.10. Interactions between processes

process A	process B	sync. condition	interaction sort	effect
g ! E1	g ! E2	value(E1) = value(E2)	value matching	synchronisation
g ! E	g ? x : t	value (E) in domain (t)	value passing	after synchronisation x=value (E)
g ? x : t	g ? y : u	t=u	value generation	after synchronisation x=y=v for v a value in domain (t)

The process is

```
process  Max 2 [a,b,c] : noexit: =
         a ? x : nat ; b ? y : nat ;
         c ! largest (x,y) ; stop
         [ ]
         b ? y : nat ; a ? x : nat ;
         c ! largest (x,y) ; stop
endprocess
```

The synchronization and communication capabilities of this process are expressed in terms of the gates a,b,c. Process Max 2 offers the choice ([ ]) between two alternative streams of actions, which only differ by the order in which the actions at gates a and b occur. Symbols "?" and "!" following a gate name denote, respectively, the input and output of some values at that gate. The input values are compared to each other and the largest appears at the output gate c [BRI-SCO-STE-87].

Behaviour Expressions

The behaviour expressions are the dynamic part of the process specification. The behaviour expressions define the syntax and semantics of LOTOS. The operators in the behaviour expressions determine the operation to be executed.

Operators

a) Stop

This keyword represents an inactive process. It cannot perform any event, and is as basic in LOTOS as the number zero is in arithmetic.

b) Action-prefix

If "a" is an event offer and "P" a process, then "a ; P" is a new process offering "a" first and behaving like "P" afterwards. The operator ";" represents a certain sequentiality.

Example

```
process  one-time-buffer [in-data,
         out-data ] : = in-data
       ; out-data
       ; stop
endprocess
```

c) Choice [ ]

Let P1 and P2 be processes. Then "P1 [ ] P2" denotes the process that either behaves like P1 or P2. The choice is made non-deterministically.

Example

```
process  EXAM [ConRes, DatReq, DisInd ] :=
                  ConRes; DatReq; stop
             [ ]  ConRes; DisInd; stop
endproc
```

where the parameters are ConRes, DatReq and DisInd.

The first event is ConRes and it can be followed either by Datreq or by DisInd.

Any finite behaviour can be described with the two basic operators action-prefix and choice and *stop*, but the result would often be a superfluously long expression.

d) Parallelism ||

Let P1 and P2 be processes. Then "P1||P2" denotes the parallel composition of P1 and P2. Concurrency is modelled by arbitary interleaving of the two processes, where synchronization is possible between them as a result of suitable event offers.

Example

```
a, b, c   event offers
P1 [.....] : = a ; b; c ; stop
P2 [......] : = a; c ; stop
P [.....]: = P1||P2
```

Figure 6.6 shows the synchronization tree of the parallel composition.

e) Enabling >>

P1 and P2 are processes and $x_1,...x_n$ are free variables in P2. Then

$$P1 >> \underline{def}\ x_1 : t_1, ...\ x_n : t_n\ \underline{in}\ P2$$

denotes the sequential order of P1 and P2. After successful termination of P1, P2 is executed, where the free variables $x_1,...,x_n$ in P2 are bound to the values passed by P1.

This enabling operator can describe the sequential composition of processes directly.

Example

```
process  Sender [ ConReq, ConCnf,
         DatReq, Disreq ] :=
         Connection-Phase  ConReq, ConCnf
         Data-Phase  [ DatReq, DisReq ]
endprocess
```

where Connection-Phase and Data-Phase are subprocesses. The Data-Phase can be executed only after the successful termination of the Connection-Phase.

f) Exit

A successfully terminating process is represented by *exit*. This can be parametrized by value expressions which are passed to the subsequent process.

Example

If we want describe the subprocesses of the example demonstrating the enabling operation, both keywords (*stop* and *exit*) are used.

```
process   Connection-Phase [ ConReq,
          ConCnf ] : = Conreq; ConCnf;
          exit
endprocess
process   Data-Phase [ DatReq, DisReq]: =
          ( DatReq; Data-Phase[DatReq,
          DisReq ] [ ] DisReq; stop
          )
endprocess
```

g) Disabling [>

Let P1 and P2 be processes. Then "P1[>P2" defines a process where at each point of the execution of P1, an initial event of P2 can occur. In such a case, the execution is transferred to P2, leaving P1.

Example

Let P1 and P2 be the following processes denoted as Activity and Disrupt, respectively. The first process is

```
 process  Activity [ a,b,c ]: =
 a ; b ; c ; Activity [ a,b,c ]
endproc
```

The disruption of the processes is described by the disabling operator
Activity [a,b,c] [>Disrupt [discon, reason]
It means that the second process has a priority and every event of the first process can be disrupted by the second. The above behaviour expression can be as follows

```
  ( discon ; reason ; stop
[ ]a ; ( discon ; reason ; stop
     [ ] b ; ( discon ; reason ; stop
        [ ]c ; ( Activity [ a,b,c ]
          [ > Disrupt [ discon, reason]
 )    )  )     )
```

h) Guarding →

Let B be a predicate or condition and P a process. Then "[B] → P" denotes the process that only behaves like P if B is fulfilled and otherwise denotes the empty behaviour.

Example

We define the first steps of connection establishment of a transport protocol (class 0). B is a predicate denoting whether the network connection is open or not. The process P describes the state transition from CLOSED to WFCC.

```
[Network Connection] → P ; exit
where process P :=
  T CONreq
; N CONreq
; CR
endprocess
```

**i) Restriction **

Let B be a behaviour and G a set of gates. Then "B\G" denotes the behaviour of F but without every interaction executed through G.

Example

```
process Protocol :=
        (Sender [ ] Receiver) \ Medium
endprocess
```

Table 7.11 summarizes the basic LOTOS operators. The full LOTOS [IS 8807] has more operators. These are necessary for advanced LOTOS programming, but the problems can also be solved with the basic set using more programming steps.

The following conventions are used in the table:

(a) B, B1, B2 are behaviour - expressions
(b) BE denotes a boolean - expression
(c) E1, ...En denote value - expressions
(d) g, g1,...gn are gate - identifiers
(e) t, t1,...tn denote sort - identifiers
(f) x,x1,...xn denote value - identifiers

Table 7.11. LOTOS syntax table with basic operators

NAME	CONCRETE SYNTAX
inaction	s t o p
termination	e x i t
choice	B1 [ ] B2
action-prefix	g;B, g?x:t;B,g!E;B,i;b
parallel-composition	B1 || B2
restriction	B \ g
guarding	[BE] -> B
disabling	B1 [> B2
enabling	B1 >> B2
sum-expression	choice x:t [ ] B

LOTOS Data Types

LOTOS applies a special language — ACT ONE — for specifying data types. ACT ONE (see Section 6.2) is a specification language for abstract data types. The abstract data types (ADT) are algebras.

The specification of an ADT contains the signature and the equations. The signature defines the syntax of a data type and is composed of sorts and operations.

E x a m p l e

```
type Enriched nat is Nat numbers with
    opns +: nat, nat -> nat
    eqns    for all x,y: nat
            of sort nat
            x + 0 = x ;
            x + succ y = succ x+y ;
endtype
```

The name of our algebra is "Enriched nat". This example is a stepwise refinement of the example on page 162, where we defined the natural numbers. The keywords *is* and *with* together help us to base a new specification on an old one. This combination is useful because predefined definitions can be placed in a library and if we call them, we can make the final specification more simple.

The *parametrization* has a special importance specifying complex data types, e.g. arrays, stacks and queues. The parametrization is a combination of a formal parameter part with the sorts, operations and equations of the parametrized type definition. The formal parameter can be actualized by different actual names of the sorts and operations.

Example

A queue will be parametrized in our example with the natural numbers and the characters respectively.

```
type  queue (data) is  data  with
        sorts   queue
        opns    create: —► queue
                add: data, queue —► queue
                first:  queue —► data
        eqns    ...
                first  create  = d0;
                ...
endtype
type  data  is
        formalsorts  data
        formalopns   d0: —► data
endtype
```

The formal parameter "data" can be actualized by the "Natural numbers" or "Characters":

```
type  Nat number queue  is
          queue (data) actualizedby
                      Nat numbers  using
          sortnames  nat  for  data
          opnames      0 for  d0
endtype
```

```
type  Character queue  is
      queue (data) actualizedby
                   Characters  using
      sortnames  char  for  data
      opnames  e  for  do
endtype
```

The above example demonstrates well how the parametrization solves the problem of the substitution of the formal parameter with different actual parameters.

The *renaming* changes type and data without any changes in the semantic. The keyword is *renamedby* and the names to be substituted are given in parentheses.

Example

```
type  connection  objects  is
      queue (data) renamedby
      sortnames  channel  for  queue
                 objects  for  data
      opnames  send  for  add
               receive  for  first
endtype
```

Recursion and Nondeterminism

Infinite behaviour can be defined using the recursive occurrence of process identifiers in behaviour expressions.

Example

```
process  receiver  [data, ack] :=
          data
         ;ack
         ;receiver [data, ack]
endprocess
```

This process receives data and acknowledgement, alternatively.

Recursion makes it easier and more concise to describe processes by regular repetitions.

Another important feature of LOTOS is its capability to handle nondeterminism. If a process is constructed of parallel subprocesses starting with the same event and finishing in some other way, the original process is nondeterministic.

E x a m p l e

```
      a ; b ; stop
  [ ] a ; c ; stop
```

The internal event i can also describe certain nondeterminism.

E x a m p l e

```
        connection - confirm
  [ ]i; disconnect - request
```

The above process generally receives connection-confirm, but sometimes a disconnect-request in consequence of an internal event.

PART III
APPLICATIONS

8. ALTERNATING-BIT PROTOCOL SPECIFICATION

8.1 STATE-TRANSITION GRAPH OF THE ALTERNATING-BIT PROTOCOL

The alternating-bit protocol is the simplest protocol with sequence numbering. Every message can be composed of three parts: [BAR-SCA-69], [LYNCH-68], [LYNCH-69] (Fig. 8.1)

— information (data or acknowledgement),
— error detection,
— control bit.

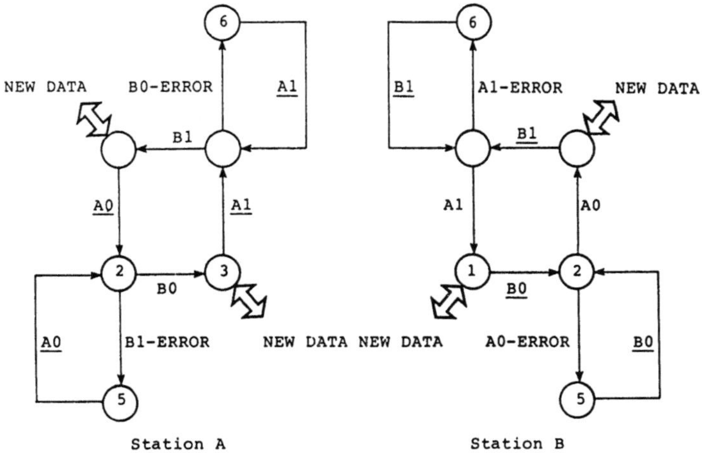

Fig. 8.1. Operation of the AB protocol

There is only a single frame-type, no control frame is needed. The single frame-type contains information as control. It is suitable for traffic between two stations therefore it is not necessary to have any addressing. The error-detecting part detects the bit-errors, the control bit determines the sequence numbering as the acknowledgement.

The control bit of the sent messages is 0 and 1 alternating, therefore it is known as the alternating-bit protocol. The control bit of the response message is always identical with the control bit of the sent message. if the other partner does not want to send an information message, he can answer with a simple acknowledgement, the control bit fulfils the function of the acknowledgement. If a response message arrives to the sender with an erroneous control bit, then the sender repeats its previous message with unchanged control bit.

The following states and events are demonstrated in the state-transition graphs (Fig. 8.2); Table 8.1 contains the interpretation of the states and events. The state-transition tables are represented in Table 8.2.

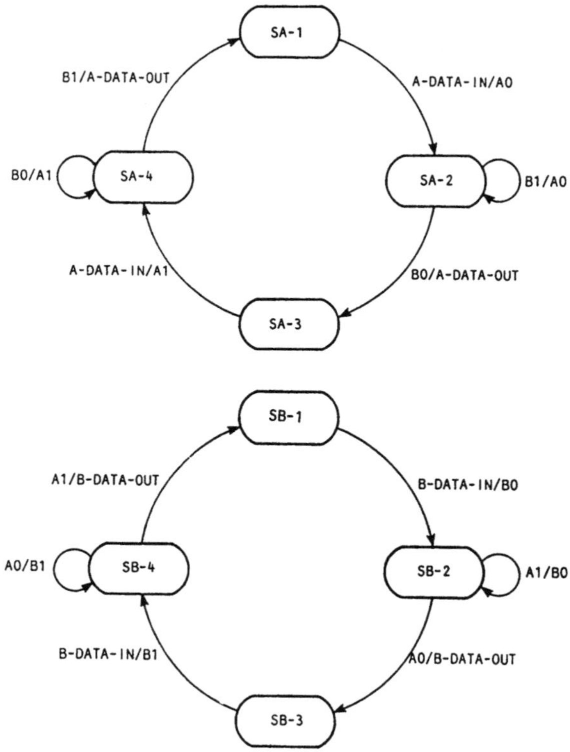

Fig. 8.2. State-transition graph of AB protocol

Table 8.1. States and events of the AB protocol

A-station

STATE	INTERPRETATION
SA-1	Response arrived with "1" control bit
SA-2	Message sent with "0" control bit
SA-3	Response arrived with "0" control bit
SA-4	Message sent with "1" control bit

INPUT EVENT	INTERPRETATION
A-DATA-IN	Input message of A-station
B_0	Message of B-station with "0" control bit
B_1	Message of B-station with "1" control bit

OUTPUT EVENT	INTERPRETATION
A-DATA-OUT	Output message of A-station
A_0	Message of A-station with "0" control bit
A_1	Message of A-station with "1" control bit

B-station

STATE	INTERPRETATION
SB-1	Response arrived with "1" control bit
SB-2	Message sent with "0" control bit
SB-3	Response arrived with "0" control bit
SB-4	Message sent with "1" control bit

Table 8.1 (continued)

INPUT EVENT	INTERPRETATION
B-DATA-IN	Input message of B-station
A_0	Message of A-station with "0" control bit
A_1	Message of A-station with "1" control bit

OUTPUT EVENT	INTERPRETATION
B-DATA-OUT	Output message of B-station
B_0	Message of B-station with "0" control bit
B_1	Message of B-station with "1" control bit

Table 8.2. State-transition tables of the AB protocol

State-transition table of A-station

State \ Input event	B_o	B_1	A-DATA-IN
SA-1	-	-	SA-2/A_o
SA-2	SA-3/A-DATA-OUT	SA-2/A_o	-
SA-3	-	-	SA-4/A_1
SA-4	SA-4/A_1	SA-1/A-DATA-OUT	-

State-transition table of B-station

State \ Input event	A_o	A_1	B-DATA-IN
SB-1	-	-	SB-2/B_o
SB-2	SB-3/B-DATA-OUT	SB-2/B_o	-
SB-3	-	-	SB-4/B_1
SB-4	SB-4/B_1	SB-1/B-DATA-OUT	-

8.2 ALTERNATING-BIT PROTOCOL DESCRIPTION WITH PETRI NET

The Petri net description gives a more microscopic picture of the operation of the alternating-bit protocol. A single net (Fig. 8.3) can modelize the sender, the receiver and the transmission medium.

The conditions sending a message with "0" control bit are described in Table 8.3. The same letters are used for the same conditions for sending a message with control bit "1", as for control bit "0". The only difference is the number "1" after the letter instead of "0".

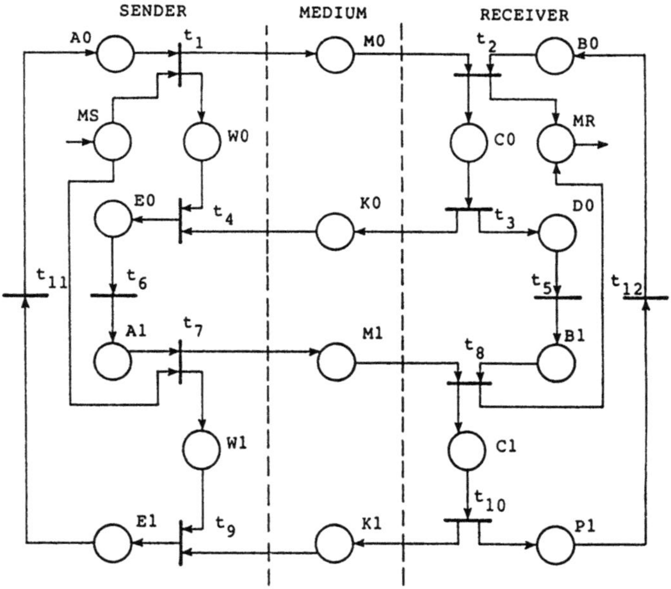

Fig. 8.3. Petri net specification of the AB protocol

Table 8.3. Conditions of a message with "0" control bit

A_o	the sender is ready to send a message with "0" control bit
B_o	the receiver is ready to receive a message with "0" control bit
M_o	transmission of a message with "0" control bit is in progress
K_o	acknowledgement of the message with "0" control bit is in progress
W_o	the sender is waiting for the acknowledgement of a message with "0" control bit
D_o	acknowledgement of a message with "0" control bit has been sent
C_o	a message with "0" control bit arrived to the receiver
E_o	a message with "0" control bit arrived to the sender

Table 8.4. Interpretation of the events

t_1	a message starts with "0" control bit
t_2	a message arrives with "0" control bit
t_3	an acknowledgement starts with "0" control bit
t_4	an acknowledgement arrives with "0" control bit
t_5	the receiver finishes the reception of a message with "0" control bit
t_6	the sender is ready to send a message with "1" control bit
t_7	a message starts with "1" control bit
t_8	a message arrives with "1" control bit
t_9	an acknowledgement arrives with "1" control bit
t_{10}	an acknowledgement starts with "1" control bit
t_{11}	the sender is ready to send a message with "0" control bit
t_{12}	the receiver finishes the reception of a message with "1" control bit

The interpretation of the events is in Table 8.4. The operation of the protocol can be followed from Fig. 8.2. Choose the initial notation having tokens in A_0 and B_0. In the first step the transition t_1 fires, the other initiating token and the token of M_0 goes to C_0, t_3 fires. If t_3 fires, then the acknowledgement of the message with "0" starts (K_0 becomes marked) and the receiver prepares itself to receive the message with control bit "1" (D_0 becomes marked). The acknowledgement of the message with "0" arrives into the sender (t_4 fires).

Then t_5 and t_6 fire and the previous process is repeated, but now concerning the message with control bit "1". The firing sequence is

$$t_7, t_8, t_{10}, t_9, t_{11}, t_{12}$$

At the end of the firing sequence the tokens return to A_0 and B_0 and the initial notation is re-established.

The model can be completed to include the erroneous cases (the control bit of the acknowledgement is the opposite), by inserting a new transition into the parts with control bit "0" and "1" (Fig. 8.4).

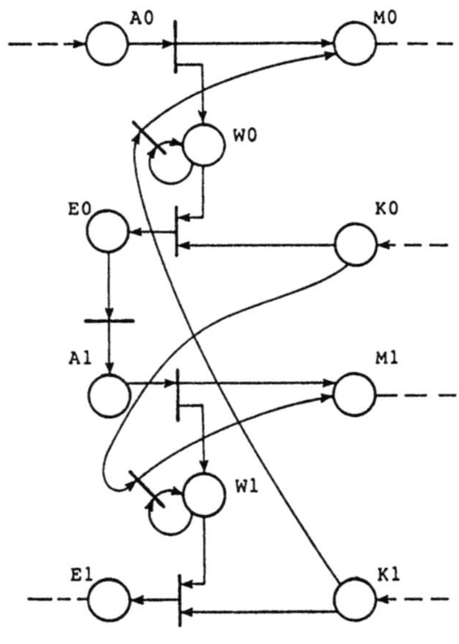

Fig. 8.4. Detailed Petri net of AB protocol

This new transition fires after the arrival of an acknowledgement with control bit "1" as a response on information with control bit "0" and the other transition fires in the opposite case. Now tokens arrive into the node of the transmission medium M and into the node of the acknowledgement W.

A Petri Net Specification Complemented with Variables

The first step towards the ESTELLE specification language was taken in [BO-GE-77]. Keller's model [KEL-76] was extended to distributed systems. Keller's original model is essentially a Petri net composed of a set of places and transitions complemented with a set of variables. Each transition in the net has associated with it an enabling predicate, depending on some variables and an action, assigning new values to some variables. This model incorporates both the state machine and programming language formalism.

The extended model for distributed systems has disjoint subsystems and some means of communication between them. For the interaction of different subsystems, each subsystem may contain certain distantly initiated actions. The initiation of an action in a distant subsystem corresponds to the sending of a message and the execution of the distantly initiated action corresponds to the receiving of the message by the distant subsystem.

[BO-GE-77] uses a notation close to the programming language Pascal for the specification of the variable declarations, predicates and actions of a subsystem.

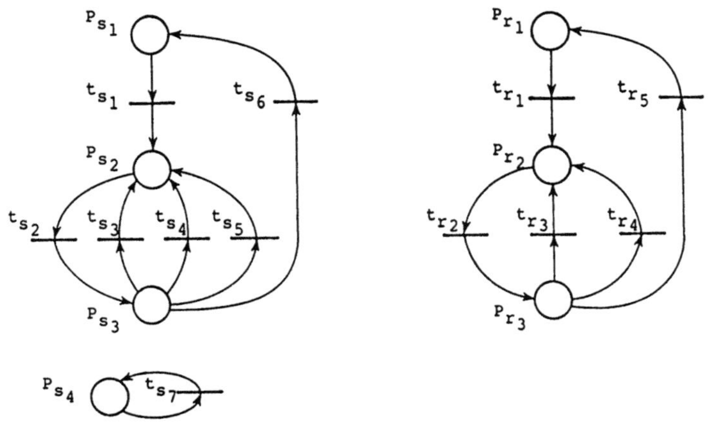

Fig. 8.5. Petri net specification model of the AB protocol

This extended model describes the alternating-bit protocol composed of a sender and a receiver (Fig. 8.5). The sender has a timer, too. The variables and actions are described in Table 8.5 and Table 8.6, respectively.

This model is a transition from a basic Petri net to an Estelle specification. The next section contains an Estelle specification of an alternating bit protocol.

Table 8.5. Variables of the AB protocol specification

```
Sender

seq: (0,1)                /* sequence number of the
                             last message
ack: (0,1,error,none)     /* acknowledgement from the
                             receiver
data: ...                 /* data to be sent
timeout: boolean          /* timeout
time: integer             /* counter

Initial State: seq = 1



Receiver

exp: (0,1)                /* expected sequence number
                             for the next message
seqnb: (0,1,error,none) /* sequence number of the
                             message received last
data: ...                 /* data received last

Initial State: exp = 1, seqnb = none
```

Table 8.6. Actions of the AB protocol specification

A C T I O N S

<u>S e n d e r</u>

Transi-tion	Enabling predicate	Action	Meaning
t_{s_1}	true	new(data); seq:=(seq+1) mod 2;	*fetch user-data* *calculate sequence number*
t_{s_2}	true	INITIATE(transD,seq, data); ack:=none; time:=t_0; timeout:=false;	*sending of the message <seq,data>*
t_{s_3}	ack=(seq+1) mod 2	;	*reception of a wrong acknowledgement*
t_{s_4}	ack=error	;	*erroneous acknowledgement*
t_{s_5}	timeout	;	*timeout*
t_{s_6}	ack=seq	;	*reception of the expected acknowledgement*

t_{s_7}	true	time:=time-1;
		if time=0 then
		timeout:=true;
			timer

Distantly initiated action	Action	Meaning

transA (p:(0,1))	case transmission of
	correct; ack:=p;
	erroneous:
	ack:=error;
	loss: ;
		
		reception of an acknowledgement
		erroneous reception
		
		loss

Table 8.6 (continued)

Receiver

Transition	Enabling predicate	Action	Meaning
t_{r_1}	true	use(data); exp:=(exp+1) mod 2;	*delivery of a received message*
t_{r_2}	true	INITIATE(transA,exp)	*sending of an acknowledgement*
		seqnb:=none;	
t_{r_3}	seqnb=exp	;	*reception of a message with a wrong sequence number*
t_{r_4}	seqnb=error	;	*erroneous message*
t_{r_5}	seqnb=(exp+1)mod2	;	*correct message*

Distantly initiated action	Action	Meaning
transD (p_1:(0,1); p_2: ...)	case transmission of	
	correct: seqnb:=p_1 data:=p_2;	*correct reception*
	erroneous: seqnb:=error;	*erroneous message*
	loss: ;	*loss*

8.3 ESTELLE SPECIFICATION OF THE ALTERNATING-BIT PROTOCOL

The Modules

A data communications protocol can be modelled in ESTELLE as one or more modules. The alternating-bit protocol is composed of the following modules (Fig. 8.6):

— user module,

— alternating-bit module;

- data transfer submodule,
- timer submodule,

— network module.

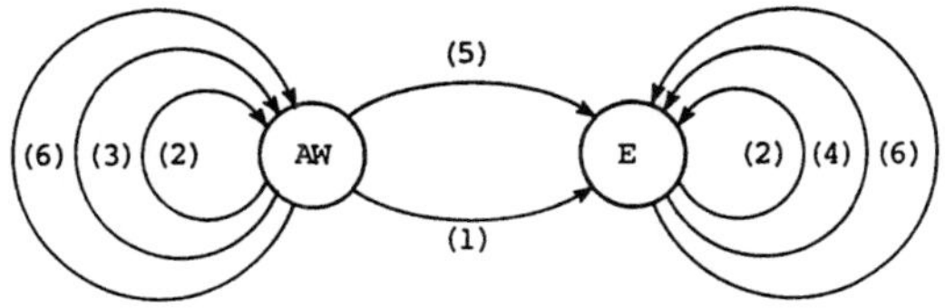


```
Explanations

(1) transition
     U.SEND_request/N.DATA_request and S.TIMER_request

(2) transition
     U.REC_request/U.REC_response

(3) transition
     S.TIMER_response/N.DATA_request and S.TIMER_request

(4) transition
     S.TIMER_response/-

(5) transition
     N.DATA_response/-

(6) transition
     N.DATA_response/N.DATA_request

States:
     E  ESTAB    Establish
     AW ACK_WAIT Acknowledge waited
```

Fig. 8.6. State-transition graph of the data transfer submodule

The user and network modules are external, process type modules. Both are connected to the alternating bit module through a channel. The channel is defined with the interaction point. The user module has an interaction point named U, its channel type is named "U-access point" and the role of the module with respect to U is named "User".

The interaction primitives are

— SEND-request,
— RECEIVE-request,
— RECEIVE-response.

The network module has an interaction point named N, its channel type is named "N-access-point".

The interaction primitives are

— DATA-request and

— DATA-response.

The main part is the alternating bit module. It has two interaction points V and N. The roles of the module are named "Provider" with respect to U and named "User" with respect to N. The alternating bit module is a parent module, its submodules are the data transfer and the timer modules.

Every module has a module header and a module body. The main components of the body are

— the declaration part,

— the initialization part,

— the transition declaration part,

— the termination part.

The declaration and the transition declaration of the data transfer submodule will be introduced as an example in the following subsections. The entire Estelle description can be found in Annex 2, Section 2.7 of the standard for Estelle and an extracted version is given in [LINN-86].

Declaration of the Data Transfer Submodule

The declaration part describes the variables, states, functions and procedures belonging to the module. Generally the variables are enumerated first:

```
var
      Send_buffer, Recv_buffer:   Buffer_type;
      Send_seq,Recv_seq:          Seq_type;
      P,Q:                        Msg_type;
      B:                          Ndata_type;
```

Different procedures and functions help to make precise the variable declarations. Easy-to-understand examples are the sequencing procedures describing the alteration rules of the zero and one bit.

```
      procedure Inc_send_seq;
        begin
            Send_seq := (Send_seq + 1) mod 2
        end;
      procedure Inc_recv_seq;
        begin
            Recv_seq := (Recv_seq + 1) mod 2
        end;
```

The alternating bit protocol has two states (ACK_WAIT, ESTAB) and in state set the notation **EITHER** is used for any state if it has no significant meaning

```
state: (ACK_WAIT, ESTAB); { enumeration of states }

stateset { state-set-definition-part }
EITHER = [ACK_WAIT, ESTAB];
```

Transitions

The data transfer submodule contains all the transitions in the alternating bit protocol. The finite state-machine version of the data transfer submodule is shown in Fig. 8.7. The number of brackets assigned to each arc is the number of the transition according to the Estelle program. The from and to keywords introduce the current state and next state, respectively. There are no priority and provided clauses in the transitions. The transition block is described between the begin and end. The transition block contains the output messages and the instantiation of the procedures.

The Estelle description of the transitions is the following:

```
trans { transition-declaration-part of
        the data-transfer-activity }
```

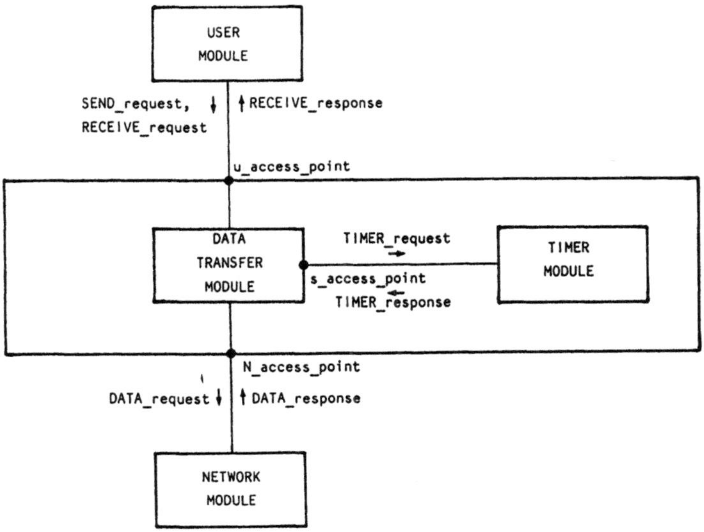

Fig. 8.7. Alternating bit protocol modules

```
from ESTAB
    to ACK_WAIT
        when U.SEND_request
         begin                          {transition 1 }
             P.Msgdata := Udata;
             P.Msgseq := Send_seq;
             Store(Send_buffer,P);
             Format_data(P,B);
             output N.DATA_request(B);
             output S.TIMER_request
         end;

from EITHER
    to same
        when U.RECEIVE_request
            provided not buffer_empty(Recv_buffer)
             begin                      {transition 2 }
             Q := Retrieve(Recv_buffer);
             output U.RECEIVE_response(Q.Msgdata);
             Remove(Recv_buffer, Q)
             end;
```

```
from ACK_WAIT
    to ACK_WAIT
        when S.TIMER_response
         begin                            { transition 3 }
             P := Retrieve(Send_buffer);
             Format_data(P,B);
             output N.DATA_request(B);
             output S. TIMER_request
         end;

from ESTAB
    to ESTAB
        when S.TIMER_response
         begin                            { transition 4 }
             { Do nothing; the message that caused this
                 timer to be set has been acknowledged. }
         end;

from ACK_WAIT
    to ESTAB
       when N.DATA_response
         provided Ack_ok(Ndata)
         begin                             { transition 5 }
             P := Retrieve(Send_buffer);
             Remove(Send_buffer, P);
             Inc_send_seq
         end;
```

```
from EITHER
    to same
       when N.DATA_response
            provided Ndata.Id = DATA
             begin                          { transition 6 }
                Q.Msgdata := Ndata.Data;
                Q.Msgseq := Ndata.Seq;
                Format_ack(Q,B);
                output N.DATA_request(B);
                if Ndata.Seq = Recv_seq then
                 begin
                      Store(Recv_buffer,Q);
                      Inc_recv_seq
                 end
            end;
end; { of the data transfer activity body  and end of the
         module-declaration-part of the Alternation_bit_body }
```

9. PROTOCOL SPECIFICATIONS IN THE DATA LINK LAYER

9.1 THE HYBRID HDLC MODEL

Bochmann [BOCH-80] described two asymmetric HDLC versions, the NRM and ARM mode. The HDLC was functionally decomposed and every subsystem has program variables and instructions. First of all the HDLC functions were described. These functions are the following:

— connection establishment and release,
— information transfer (source, sink, transmission and check-point),
— PF control,
— timing control (clock).

The relation of the functions is demonstrated in Fig. 9.1. The source and sink are self contained modules, the transmission and check-point modules have no state-transitions, only procedures activated by other modules.

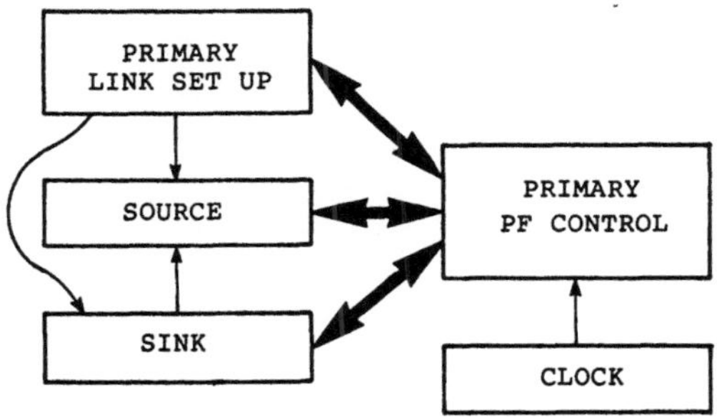

Fig. 9.1. Modules of an HDLC primary station

The source and sink modules depend hierarchically on the link-set-up module of the primary, data exchange between the source and sink can be realized only in connected state of the link-set-up module.

The common variable of the source and sink is the acknowledgement delivered by the information frame. The common variable of the clock and the PF control is the time-out condition.

The PF control module is directly coupled to the link-set-up module as to well as the sink and source. The sending and receiving state transitions of every three modules are coordinated by the state transition of the PF control module, when it controls the validity of the sent and received PF bits.

Every functional element is characterized by

— states,
— state transitions,
— variables,
— enabling predicates,
— actions.

The state-transition graphs of the *link-set-up* module are demonstrated in Fig. 9.2.

Link set up

Primary

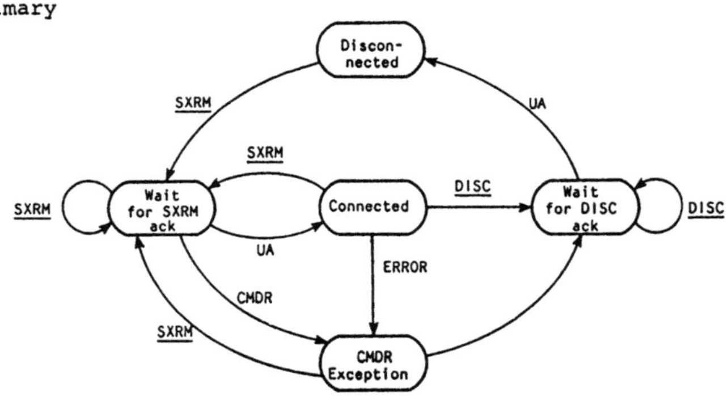

Secondary

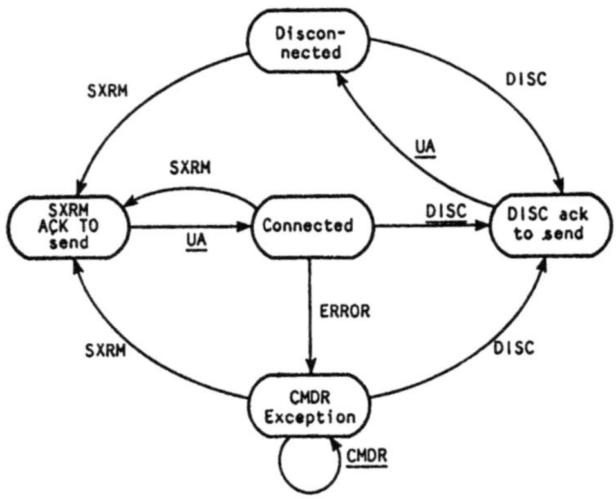

Fig. 9.2. State-transition graphs of the link-set-up module

Table 9.1. Link-set-up components

Transition	Enabling predicate	Action	Meaning
Primary station:			
SXRM	PF-control.bit = 1	send-unnumbered (SXRM) ;	SXRM is SNRM or SARM depending on the mode to be set
UA	received.kind = UA	init (source) ; init (sink) ; init (transmission) ;	initialize the source and sink components
DISC	PF-control.bit = 1	send-unnumbered (DISC) ;	
CMDR	received.kind = CMDR	init (transmission) ;	
ERROR	Status in [invalid-control-field, invalid-info, invalid-size, invalid-NR]	init (transmission) ;	frame received contained an error to be resolved by a higher level recovery procedure at Primary
OTHER	...	init (transmission) ;	in certain states, the reception of certain kinds of frames is simply ignored (not shown in the transition diagram).
Secondary station:			
SXRM	received.kind = SXRM	init (transmission) ;	SXRM is SNRM or SARM depending on the mode set
UA	PF-control.bit = 1	send-unnumbered (UA) ;	
DISC	received.kind = DISC	init (transmission) ;	
CMDR	true	send-CMDR (received,control-field, source. VS, sink. VR, status) ;	report procedural error for recovery by primary
ERROR (as above)			
NOT SXRM DISC	not (received.kind in [SNRM, SARM, DISC])	init (transmission) ;	

The states of the primary station are

— disconnected,
— wait for SXRM ack,
— connected,
— wait for DISC ack,
— CMDR exception.

SXRM represents either SNRM or SARM depending on the mode to be set.

The input and output events of the primary are: *SXRM*, *DISC*, *UA*, *CMDR*, *ERROR*. The underlined events are output messages, the not-underlined ones are input messages.

The initial state of the primary station is disconnected.

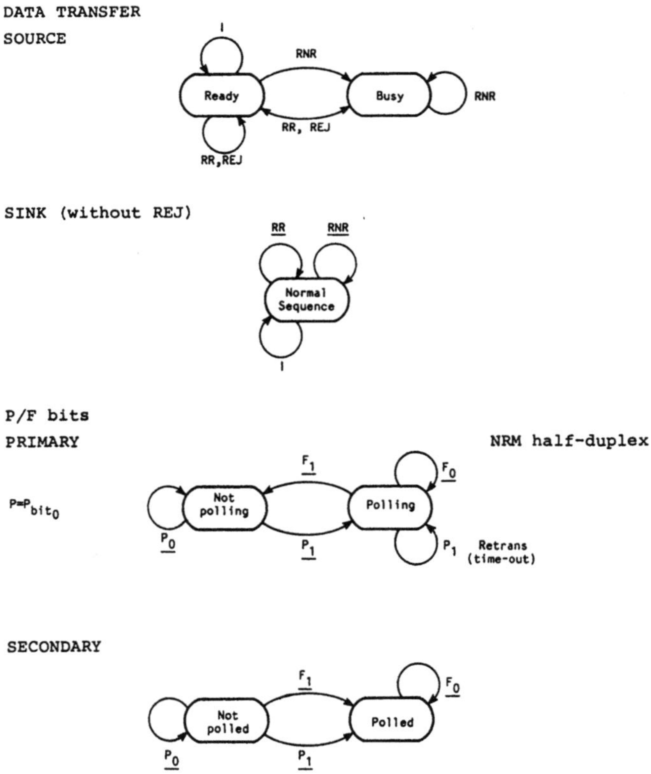

Fig. 9.3. Data transfer

The states of the secondary station are:
— disconnected,
— SXRM ack to send,
— connected,
— DISC ack to send,
— CMDR exception.

The state transitions of the secondary station are: SXRM, DISC, *UA*, *CMDR*.

Table 9.1 enumerates the state transitions, the enabling predicates, and the actions of the link-set-up module.

Table 9.2. Variables of the source and sink module

SOURCE MODULE

VARIABLES	MEANING
- unack: sequence count	oldest frame sent in sequence but not yet acknowledged
- VS: sequence count	sending state variable, the sequence number of the next frame to be sent
- buffer	
- data (VS: sequence count)	function which returns the buffer of data to be sent
- to-send (Vs: sequence count)	Boolean function which is true when there is a buffer of data to be sent
- free-until (NR: sequence number)	procedure to release buffers of data after the arrival of the acknowledgement of NR; the receiving sequence number NR is the expected sequence number of the next I-frame

SINK MODULE

VARIABLES	MEANING
- VR: sequence count	sequence number of the next I-frame
- buffer	
space	Boolean function which returns the value true when there is a buffer available for receiving data

A detailed description of every module is in [BOCH-77]. The state transitions can be seen in Figs. 9.2 and 9.3. As an illustration, the variables of the source and sink module are given in Table 9.2.

This model is an additional version of the unified model of Bochmann and Gecsei [BOCH-GE-77]. Its power is in terms which are easy to use in a high-level programming language, such as Pascal, and in program variables. This version is relatively easy to implement and this was the first step towards the ESTELLE language [BOCH-78].

9.2 A FORMAL GRAMMAR DESCRIPTION OF HDLC

Essence of the Method

The formal grammar description for protocols was first suggested by Harangozó [HAR-77]. His formal description method based on formal languages considers the protocol as a dialogue consisting of sentences. The sentences characterize the communication and the interaction between the remote partners or peer entities.

The sequence of sentences shows an ordered structure so it can be supposed that they might be generated by a grammar. If we can find the sets of symbols from which the participants compose their sent sentences and we can construct the grammar containing the rules correctly generating these sentences, we can get the formal description of the protocol.

The steps for constructing such a grammar are:

— to determine the set of symbols sent by the first partner to the second and to determine the set of symbols sent by the second partner to the first. The union of these two sets forms the set of symbols characterizing the message exchange, that can be sent by the protocol machine, so they will form *the set of terminals*;

— to provide the *non-terminals*;

— to construct the set of *production rules* for the grammar following the sequence of the dialogue.

In the following sections we attempt to construct some models of the HDLC procedure. The models reflect the characteristics of the dialogue in half-duplex and full-duplex mode. A number of other improved models are mentioned taking into consideration the effect of time-out, unknown frame, sequence numbering and the request for retransmission frames with a sequence number. Finally the advantages and limitations of this approach are emphasized [HAR-76], [HAR-78a], [HAR-78b].

Half-duplex Models

Let the basic grammar describing the dialogue be

$$G = (V_T, V_N, P, S)$$

the elements of which can be constructed in the following way:

$$V_T = V_{T1} \cup V_{T2} ,$$

where

V_{T1} is the set of symbols transmitted by the primary, and
V_{T2} is the set of symbols transmitted by the secondary station.

The production rules are the following:

$$P = \{W \rightarrow aX\} \cup \{W \rightarrow a\} \cup \{Y \rightarrow bZ\} \cup \{Y \rightarrow b\}$$

where

$W, X, Y, Z \in V_N$;
$a \in V_{T1}$; $b \in V_{T2}$.
V_N is given after the construction of production rules.

Considering the concrete grammar model of the HDLC procedure in HDX and NRM let us define the terminals.

$$V_{T1} = \{\text{snrm, disc, i, rr, rnr}\} ;$$
$$V_{T2} = \{\text{ua, cmdr, i, rr, rnr}\} ;$$

$\rightarrow$ denotes the direction of transmission from the primary to the secondary;
$\leftarrow$ denotes the direction of transmission from the secondary to the primary.

The concrete production rules are

$P = \{S \rightarrow \underrightarrow{\text{snrm}}\ A$	$E \rightarrow \underrightarrow{\text{rr}}\ D$
$A \rightarrow \underleftarrow{\text{ua}}\ B$	$E \rightarrow \underrightarrow{\text{rnr}}\ K$
$A \rightarrow \underleftarrow{\text{cmdr}}$	$E \rightarrow \underrightarrow{\text{disc}}\ H$
$B \rightarrow \underrightarrow{\text{i}}\ C$	$F \rightarrow \underleftarrow{\text{i}}\ E$
$B \rightarrow \underrightarrow{\text{rr}}\ D$	$F \rightarrow \underleftarrow{\text{rr}}\ B$
$B \rightarrow \underrightarrow{\text{disc}}\ H$	$F \rightarrow \underleftarrow{\text{cmdr}}$
$C \rightarrow \underrightarrow{\text{i}}\ C$	$G \rightarrow \underrightarrow{\text{i}}\ C$
$C \rightarrow \underleftarrow{\text{i}}\ E$	$G \rightarrow \underrightarrow{\text{rr}}\ D$
$C \rightarrow \underleftarrow{\text{rr}}\ B$	$G \rightarrow \underrightarrow{\text{snrm}}\ A$
$C \rightarrow \underleftarrow{\text{rnr}}\ J$	$G \rightarrow \underrightarrow{\text{disc}}\ H$
$C \rightarrow \underleftarrow{\text{cmdr}}$	$H \rightarrow \underleftarrow{\text{ua}}$
$D \rightarrow \underleftarrow{\text{i}}\ E$	$H \rightarrow \underleftarrow{\text{cmdr}}$
$D \rightarrow \underleftarrow{\text{rr}}\ B$	$J \rightarrow \underrightarrow{\text{rr}}\ F$
$D \rightarrow \underleftarrow{\text{cmdr}}$	$J \rightarrow \underrightarrow{\text{disc}}\ H$
$E \rightarrow \underrightarrow{\text{i}}\ C$	$K \rightarrow \underleftarrow{\text{rr}}\ G$
$E \rightarrow \underleftarrow{\text{i}}\ E$	$K \rightarrow \underleftarrow{\text{cmdr}}\}$;

The non-terminals are the following

$$V_N = \{S, A, B, C, D, E, F, G, H, J, K\} .$$

The equivalent automaton can be constructed and the state transition graph of the automaton helps us to follow the steps of the dialogue.

This basic model can be improved to describe other effects (e.g. time-out, unknown frame, sequence numbering). The improved model for time-out and unknown frame has the same principles. The set of terminals is increased with the symbol describing either the time-out or the unknown frame and the set of non-terminals is increased with the set of additional non-terminals necessary for the new production rules.

The terminals for the *time-out* case are

$$V_T^1 = V_{T1} \cup V_{T2} \cup \{to\} ,$$

where

to is a terminal symbol characterizing the time-out effect taken into account at the primary station.

The non-terminals are

$$V_N^1 = V_N \cup V_{NR}^1$$

where

V_{NR}^1 is the set of additional non-terminals which is given by the production rules taking into account the effect of time-out.

The improved model for unknown frame is similar to the time-out model.

Improved Model with Sequence Numbering

The essence of the improvement is to provide the terminal symbols representing the information (**I**)- and supervisory (**S**)-frames with indexes according to the sequence number of frames. The production rules need to be modified too, in order to provide the production of new indexes. These modifications enlarge the set of non-terminals and the non-terminals help in the production of new indexes of the terminals.

Let us use the following indexes for the **I**- and **S**-frames.

$$i_{pq}, rr_q, rnr_q \in V_{T1} ,$$

where

p is the sequence number of an I-command transmitted by the primary station (p = 0, 1, 2, ..., 7);

q is the sequence number of the next expected I-response by the primary station (q = 0, 1, 2, . . . , 7) .

$$i_{ts}, rr_s, rnr_s \in V_{T2}$$

where

t is the sequence number of the I-response transmitted by the secondary station ($t = 0, 1, 2, \ldots, 7$) ;

s is the sequence number of the next expected I-command by the secondary station ($s = 0, 1, 2, \ldots, 7$).

Let us construct *the grammar* of the improved model on the basis of the indexing mentioned above.

$$G^3 = (V_N^3, V_T^3, P^3, S)$$

$$V_T^3 = (V_{T1} \setminus \{\underrightarrow{i}, \underrightarrow{rr}, \underrightarrow{rnr}\}) \cup (V_{T2} \setminus \{\underleftarrow{i}, \underleftarrow{rr}, \underleftarrow{rnr}\})$$

$$\bigcup_p \bigcup_q \{\underrightarrow{i}_{pq}\} \quad \bigcup_s \bigcup_t \{\underleftarrow{i}_{ts}\}$$

$$\bigcup_q \{\underrightarrow{rr}_q, \underrightarrow{rnr}_q\} \quad \bigcup_s \{\underleftarrow{rr}_s, \underleftarrow{rnr}_s\}$$

The indexing rules are based on the following principle:

- an $\underrightarrow{i}_{pq}$ command is followed either by an $\underleftarrow{i}_{ts}$, or $\underleftarrow{rr}_s$ or $\underleftarrow{rnr}_s$ response, where $t = q$ and $s = (p+1)$;

- an $\underleftarrow{i}_{ts}$ response is followed either by an $\underrightarrow{i}_{pq}$ or $\underrightarrow{rr}_q$ or $\underrightarrow{rnr}_q$ command, where $p = s$ and $q = (t+1)$.

The production rules of the grammar of the improved model follow — taking into consideration the indexing rules.

$$
\begin{array}{ll}
P^3 = \{S \rightarrow \underrightarrow{\text{snrm}}\ A & F_{pq} \rightarrow \underrightarrow{\text{disc}}\ H \\
\quad A \rightarrow \underleftarrow{\text{ua}}\ B_{pq}\ (p=q=0) & F_{pq} \rightarrow \underleftarrow{\text{i}}_{qp}\ E_{p(q+1)} \\
\quad B_{pq} \rightarrow \underrightarrow{\text{i}}_{pq}\ C_{(p+1)q} & F_{pq} \rightarrow \underleftarrow{\text{rr}}_{p}\ B_{pq} \\
\quad B_{pq} \rightarrow \underrightarrow{\text{rr}}_{q}\ D_{pq} & F_{pq} \rightarrow \underleftarrow{\text{cmdr}} \\
\quad B_{pq} \rightarrow \underrightarrow{\text{disc}}\ H & G_{pq} \rightarrow \underrightarrow{\text{i}}_{pq}\ C_{(p+1)q} \\
\quad C_{pq} \rightarrow \underrightarrow{\text{i}}_{pq}\ C_{(p+1)q} & G_{pq} \rightarrow \underrightarrow{\text{rr}}_{q}\ D_{pq} \\
\quad C_{pq} \rightarrow \underleftarrow{\text{i}}_{qp}\ E_{p(q+1)} & G_{pq} \rightarrow \underrightarrow{\text{disc}}\ H \\
\quad C_{pq} \rightarrow \underleftarrow{\text{rr}}_{p}\ B_{pq} & G_{pq} \rightarrow \underrightarrow{\text{snrm}}\ A \\
\quad C_{pq} \rightarrow \underleftarrow{\text{rnr}}_{p}\ J_{pq} & H \rightarrow \underleftarrow{\text{ua}} \\
\quad C_{pq} \rightarrow \underleftarrow{\text{cmdr}} & H \rightarrow \underleftarrow{\text{cmdr}} \\
\quad D_{pq} \rightarrow \underleftarrow{\text{i}}_{qp}\ E_{p(q+1)} & J_{pq} \rightarrow \underrightarrow{\text{rr}}_{q}\ F_{pq} \\
\quad D_{pq} \rightarrow \underleftarrow{\text{rr}}_{p}\ B_{pq} & J_{pq} \rightarrow \underrightarrow{\text{disc}}\ H \\
\quad D_{pq} \rightarrow \underleftarrow{\text{cmdr}} & K_{pq} \rightarrow \underleftarrow{\text{rr}}_{p}\ G_{pq} \\
\quad E_{pq} \rightarrow \underrightarrow{\text{i}}_{pq}\ C_{(p+1)q} & K_{pq} \rightarrow \underleftarrow{\text{cmdr}}\ \} \\
\quad E_{pq} \rightarrow \underleftarrow{\text{i}}_{qp}\ E_{p(q+1)} & \\
\quad E_{pq} \rightarrow \underrightarrow{\text{rr}}_{q}\ D_{pq} & \\
\quad E_{pq} \rightarrow \underrightarrow{\text{rnr}}_{q}\ K_{pq}\ ; &
\end{array}
$$

$$
V_N^3 = \{S,\ A,\ H\}\ \bigcup_p \bigcup_q \{B_{pq},\ C_{pq},\ D_{pq},\ E_{pq},\ F_{pq},\ G_{pq},\ J_{pq},\ K_{pq}\}
$$

Taking into consideration the indexing rules, V_T^3 can be simplified:

$$
V_T^3 = (V_{T1} \setminus \{\underrightarrow{\text{i}},\ \underrightarrow{\text{rr}},\ \underrightarrow{\text{rnr}}\}) \cup (V_{T2} \setminus \{\underleftarrow{\text{i}},\ \underleftarrow{\text{rr}}\ \underleftarrow{\text{rnr}}\})
$$

$$
\bigcup_p \bigcup_q \{\underrightarrow{\text{i}}_{pq},\ \underleftarrow{\text{i}}_{qp}\}\ \bigcup_q \{\underrightarrow{\text{rr}}_{q},\ \underrightarrow{\text{rnr}}_{q}\}\ \bigcup_p \{\underleftarrow{\text{rr}}_{q},\ \underleftarrow{\text{rnr}}_{p}\}
$$

The production rules can be described in table form. The terminals are in the header of columns and the non-terminals determine the rows. The square determines the next non-terminal. If the non-terminal is $V_N(i)$ and the terminal is $V_T(j)$ then the square(ij) contains the next non-terminal, e.g. $V_N(m)$, where i and m are the sequence numbers of the rows and j is the sequence number of the columns.

Table 9.3 shows the production rules in table form for the basic model. The illustrative power of the table helps us to follow the entire dialogue and the table is also useful for simulating and programming the operation of the protocols.

Table 9.3. Production rules for the basic model

Non-terminals	Terminals									
	$\underrightarrow{\text{snrm}}$	$\underrightarrow{\text{disc}}$	$\underrightarrow{\text{i}}$	$\underrightarrow{\text{rr}}$	$\underrightarrow{\text{rnr}}$	$\underleftarrow{\text{ua}}$	$\underleftarrow{\text{cmdr}}$	$\underleftarrow{\text{i}}$	$\underleftarrow{\text{rr}}$	$\underleftarrow{\text{rnr}}$
S	A	-	-	-	-	-	-	-	-	-
A	-	-	-	-	-	B	A	-	-	-
B	-	H	C	D	-	-	-	-	-	-
C	-	-	C	-	-	-	C	E	B	J
D	-	-	-	-	-	-	D	E	B	-
E	-	H	C	D	K	-	-	E	-	-
F	-	-	-	-	-	-	F	E	B	-
G	A	H	C	D	-	-	-	-	-	-
H	-	-	-	-	-	H	H	-	-	-
J	-	H	-	F	-	-	-	-	-	-
K	-	-	-	-	-	-	K	-	G	-

The production rules for the improved model with indexing can be seen in Table 9.4.

Table 9.5 is an enhanced alternative of the previous grammar model taking into account the sequence numbering. The enhancement means that terminals may arrive which are not characteristic for the normal operation and the language should have new non-terminals and new production rules. For example, the connection is not established and an I-frame is sent. In this case L^3_{pq} is the new non-terminal. M^3_{pq} and N^3_{pq} are the new non-terminals, when an I-frame is sent after receiving an RNR.

Table 9.4. Production rules for grammar G^3

Non-terminals	Terminals									
	snrm	disc	i_{pq}	rr_q	rnr_q	ua	cmdr	i_{qp}	rr_p	rnr_p
S^3	A^3_{pq} p=q=0	-	-	-	-	-	-	-	-	-
A^3_{pq}	A^3_{pq} p=q=0	-	-	-	-	B^3_{pq}	C^3_{pq}	$L^3_{p(q+1)}$	L^3_{pq}	L^3_{pq}
B^3_{pq}	A^3_{pq} p=q=0	O^3_{pq}	$D^3_{(p+1)q}$	E^3_{pq}	F^3_{pq}	B^3_{pq}	-	-	-	-
C^3_{pq}	A^3_{pq} p=q=0	O^3_{pq}	-	-	-	-	C^3_{pq}	-	-	-
D^3_{pq}	-	-	$D^3_{(p+1)q}$	E^3_{pq}	F^3_{pq}	L^3_{pq}	C^3_{pq}	$H^3_{p(q+1)}$	J^3_{pq}	K^3_{pq}
E^3_{pq}	-	-	$D^3_{(p+1)q}$	E^3_{pq}	-	L^3_{pq}	C^3_{pq}	$H^3_{p(q+1)}$	J^3_{pq}	K^3_{pq}
F^3_{pq}	-	-	$D^3_{(p+1)q}$	-	F^3_{pq}	L^3_{pq}	C^3_{pq}	$*M^3_{p(q+1)}$	J^3_{pq}	$*M^3_{pq}$
O^3_{pq}	-	O^3_{pq}	-	-	-	ϕ	C^3_{pq}	$L^3_{p(q+1)}$	L^3_{pq}	L^3_{pq}
H^3_{pq}	A^3_{pq} p=q=0	O^3_{pq}	$D^3_{(p+1)q}$	E^3_{pq}	F^3_{pq}	-	-	$H^3_{p(q+1)}$	J^3_{pq}	K^3_{pq}
J^3_{pq}	A^3_{pq} p=q=0	O^3_{pq}	$D^3_{(p+1)q}$	E^3_{pq}	F^3_{pq}	-	-	$H^3_{p(q+1)}$	J^3_{pq}	-
K^3_{pq}	A^3_{pq} p=q=0	O^3_{pq}	$*N^3_{(p+1)q}$	E^3_{pq}	$*N^3_{pq}$	-	-	$H^3_{p(q+1)}$	-	K^3_{pq}
L^3_{pq}	A^3_{pq} p=q=0	O^3_{pq}	-	-	-	L^3_{pq}	L^3_{pq}	$L^3_{p(q+1)}$	L^3_{pq}	L^3_{pq}
$*M^3_{pq}$	A^3_{pq} p=q=0	O^3_{pq}	$D^3_{(p+1)q}$	$*N^3_{pq}$	$*N^3_{pq}$	-	-	$*M^3_{p(q+1)}$	$*M^3_{pq}$	$*M^3_{pq}$
$*N^3_{pq}$	-	-	$*N^3_{(p+1)q}$	$*N^3_{pq}$	$*N^3_{pq}$	L^3_{pq}	C^3_{pq}	$H^3_{p(q+1)}$	$*M^3_{pq}$	$*M^3_{pq}$

The symbol * in the table means that the incrementation of indexes is not taken into consideration at the partner side.

Table 9.5. Production rules of the improved model with indexing

Non-terminals	Terminals									
	$\underrightarrow{\text{snrm}}$	$\underrightarrow{\text{disc}}$	$\underrightarrow{\text{i}}_{pq}$	$\underrightarrow{\text{rr}}_{q}$	$\underrightarrow{\text{rnr}}_{q}$	$\underleftarrow{\text{ua}}$	$\underleftarrow{\text{cmdr}}$	$\underleftarrow{\text{i}}_{qp}$	$\underleftarrow{\text{rr}}_{p}$	$\underleftarrow{\text{rnr}}_{p}$
S	A	-	-	-	-	-	-	-	-	-
A	-	-	-	-	-	B_{pq} $p=q=0$	-	-	-	-
B_{pq}	-	H	$C_{(p+1)q}$	D_{pq}	-	-	-	-	-	-
C_{pq}	-	-	$C_{(p+1)q}$	-	-	-	C_{pq}	$E_{p(q+1)}$	B_{pq}	J_{pq}
D_{pq}	-	-	-	-	-	-	D_{pq}	$E_{p(q+1)}$	B_{pq}	-
E_{pq}	-	-	$C_{(p+1)q}$	D_{pq}	K_{pq}	-	-	$E_{p(q+1)}$	-	-
F_{pq}	-	H	-	-	-	-	F_{pq}	$E_{p(q+1)}$	B_{pq}	-
G_{pq}	A	H	$C_{(p+1)q}$	D_{pq}	-	-	-	-	-	-
H	-	-	-	-	-	H	H	-	-	-
J_{pq}	-	H	-	F_{pq}	-	-	-	-	-	-
K_{pq}	-	-	-	-	-	-	K_{pq}	-	G_{pq}	-

Full-duplex Models

The full-duplex model describes the HDLC in asynchronous response mode (ARM). The basic model is composed of two different grammars.

Let us suppose that the frames (symbols) running from the primary to the secondary are generated by a *grammar* G_1, the frames (symbols) running from the secondary to the primary are generated by a *grammar* G_2. The two transmission channels having been observed at a definite moment in time, it is possible to find only a frame (symbol) of language L_1, only a frame (symbol) of language L_2, or frames (symbols) of both languages L_1 and L_2. Consequently *grammar* G has to possess the characteristics of both grammars G_1 and G_2 to produce sentences of both languages L_1 and L_2 and, moreover, such sentences whose first part is taken from language L_1 and the second part from language L_2.

Let the two *component grammars* be

$$G_1 = (V_{N1}, V_{T1}, P_1, S_2) ,$$

$$G_2 = (V_{N2}, V_{T2}, P_2, S_2) .$$

We have to construct the resultant grammar G from the components on the basis of our method taking into consideration the conditions mentioned above.

$$G = (V_N, V_T, P, S) ;$$

$$V_N = V_{N1} \cup V_{N2} \cup \{S\} \cup V_{NF} ,$$

where

V_{NF} is the set of non-terminals which would be final states in the equivalent acceptor automaton.

$$V_T = V_{T1} \cup V_{T2} ;$$

$$P = (P_1 \setminus (\{S_1 \to \varepsilon\}, \{R \to d\})) \cup$$
$$\cup (P_2 \setminus \{S_2 \to \varepsilon\}) \cup \{S \to S_1\} \cup$$
$$\cup \{X \to W\} \cup \{Y \to Z\} \cup \{R \to d\ T\} ,$$

where

$$X \in V_{N1} \setminus S_1 ; \quad W \in V_{N2} ;$$
$$Y \in V_{N2} \setminus S_2 ; \quad Z \in V_{N1} ;$$
$$R \in V_{N1} ; \quad T \in V_{NF} ;$$
$$d \in V_{T1} .$$

When grammars G_1 and G_2 possess characteristics for generating strings of the same symbols, the sentences generated by grammar G are elements of the set $(V_{T1} \cup V_{T2})^*$.

When grammars G_1 and G_2 generate only single symbols, the sentences generated by grammar G are elements of the set $(V_{T1} \cup V_{T2} \cup V_{T1}V_{T2})$.

With regard to the conditions mentioned above, let us construct grammar G in FDX and ARM.

Let us first describe the operation of the primary with grammar G_1:

$$G_1 = (V_{N1}, V_{T1}, P_1, S_1) ,$$

where

$$V_{T1} = (\underrightarrow{\text{sarm}},\ \underrightarrow{\text{disc}},\ \underrightarrow{\text{i}},\ \underrightarrow{\text{rr}},\ \underrightarrow{\text{rnr}}) \quad ;$$

$$\begin{array}{llll} P_1 = \{S_1 \rightarrow \underrightarrow{\text{sarm}}\ A & & B \rightarrow \underrightarrow{\text{disc}} \\ \quad A \rightarrow \underrightarrow{\text{i}}\ B & & C \rightarrow \underrightarrow{\text{i}}\ B \\ \quad A \rightarrow \underrightarrow{\text{rr}}\ C & & C \rightarrow \underrightarrow{\text{rr}}\ C \\ \quad A \rightarrow \underrightarrow{\text{disc}} & & C \rightarrow \underrightarrow{\text{disc}} \\ \quad B \rightarrow \underrightarrow{\text{i}}\ B & & D \rightarrow \underrightarrow{\text{rr}}\ C \\ \quad B \rightarrow \underrightarrow{\text{rr}}\ C & & D \rightarrow \underrightarrow{\text{disc}}\} \quad ; \\ \quad B \rightarrow \underrightarrow{\text{rnr}}\ C & & \end{array}$$

$$V_{N1} = \{S_1, A, B, C, D\} \quad .$$

Let us then describe the operation of the secondary with grammar G_2:

$$G_2 = (V_{N2}, V_{T2}, P_2, S_2)\ ,$$

where

$$V_{T2} = \{\underleftarrow{\text{ua}},\ \underleftarrow{\text{cmdr}},\ \underleftarrow{\text{i}},\ \underleftarrow{\text{rr}},\ \underleftarrow{\text{rnr}}\} \quad ;$$

$$\begin{array}{lll} P_2 = \{S_2 \rightarrow \underleftarrow{\text{ua}}\ K & L \rightarrow \underleftarrow{\text{rr}}\ M & M \rightarrow \underleftarrow{\text{ua}} \\ \quad S_2 \rightarrow \underleftarrow{\text{cmdr}} & L \rightarrow \underleftarrow{\text{rnr}}\ N & M \rightarrow \underleftarrow{\text{cmdr}} \\ \quad K \rightarrow \underleftarrow{\text{i}}\ L & L \rightarrow \underleftarrow{\text{cmdr}} & N \rightarrow \underleftarrow{\text{i}}\ L \\ \quad K \rightarrow \underleftarrow{\text{rr}}\ M & M \rightarrow \underleftarrow{\text{i}}\ L & N \rightarrow \underleftarrow{\text{rr}}\ M \\ \quad L \rightarrow \underleftarrow{\text{i}}\ L & M \rightarrow \underleftarrow{\text{rr}}\ M & N \rightarrow \underleftarrow{\text{cmdr}}\} \quad ; \end{array}$$

$$V_{N2} = \{S_2, K, L, M, N\} \quad .$$

Grammar G generating the dialogue can be produced on the basis of the well-known conditions:

$$G = (V_N, V_T, P, S)$$

$$V_N = V_{N1} \cup V_{N2} \cup \{S\} \cup V_{NF} \, ,$$

where

$$V_{NF} = \{E\} \quad ,$$

$$V_N = \{S, S_1, S_2, A, B, C, D, E, K, L, M, N\} \quad ;$$

$$V_T = V_{T1} \cup V_{T2} = \{\underrightarrow{\mathrm{sarm}}, \underrightarrow{\mathrm{disc}}, \underrightarrow{\mathrm{i}}, \underrightarrow{\mathrm{rr}}, \underrightarrow{\mathrm{rnr}}, \underleftarrow{\mathrm{ua}}, \underleftarrow{\mathrm{cmdr}}, \underleftarrow{\mathrm{i}}, \underleftarrow{\mathrm{rr}}, \underleftarrow{\mathrm{rnr}}\} \quad ;$$

$$\begin{aligned}
P = {} & (P_1 \setminus (A \to \underrightarrow{\mathrm{disc}})) \cup P_2 \cup \{A \to \underrightarrow{\mathrm{disc}}\, E\} \cup \\
& \cup \{S \to S_1\} \cup \\
& \cup \{(V_{N1} \setminus \{S_1, A\}) \to (V_{N2} \setminus \{S_2, K\})\} \cup \\
& \cup \{A \to S_2\} \cup \{(V_{N2} \setminus \{S_2, K\}) \to \\
& \to (V_{N1} \setminus \{S_1, A\})\} \cup \{K \to A\} \cup \{N \to S_1\} = \\
& = (P_1 \setminus \{A \to \underrightarrow{\mathrm{disc}}\}) \cup P_2 \cup \{A \to \underrightarrow{\mathrm{disc}}\, E\} \cup \\
& \cup \{S \to S_1\} \cup \{\{B, C, D\} \to \{L, M, N\}\} \cup \\
& \cup \{A \to S_2\} \cup \{\{L, M, N\} \to \{B, C, D\}\} \cup \\
& \cup \{K \to A\} \cup \{N \to S_1\} \quad .
\end{aligned}$$

This basic model can be improved to describe
— the effect of the time-out,
— the effect of an unknown frame,
— the sequence numbering,
— the request for retransmission of frames with a sequence number.

Conclusions

The regular grammars are suitable for protocol specification in order to specify, to verify, to implement and to test the protocol, but all limitations of formal grammars are inherited by the protocol models.

The advantages of this approach are the following:

a) These models reflect the general characteristics of protocols well and they provide a logical structure supporting the deep understanding of the protocol operation and dialogues.
b) The essence of the method is strongly related to programming. A generative grammar can easily be converted to a protocol generator.
c) The basic model can be enhanced in two directions:
   — any single effect (e.g. time-out, sequencing) can be described well and can be built in the model:
   — the *side-effects* of the adjacent layers can be considered and a hierarchical protocol system can be modelled, too.

The limitations of these models are the following:

a) Protocols are examined only from structural points of view.
b) The model contains time sequences but they determine only a temporal ordering and not a time scale.
c) The protocol parameters cannot be described in the basic models, only after some model transformation.
d) The models can be enhanced and improved, but they seem to be too complex for the formal description of *real-world* protocols.
e) These concrete models do not take into consideration the lower and upper limits of the sequence numbers.

9.3 FORMAL DESCRIPTION OF LAPB WITH SDL/GR

The Structure of LAPB Description

The CCITT developed the SDL language for the formulation and solution of communication tasks. In particular, the Australian Special Group did very much to describe the protocols with SDL graphs. The structure of this LAPB description to be discussed now was worked out by Dickson and Wheeler of the Telecom Australia Research Laboratories [WHE-CHI-84].

The objective of the description is the definition of a data-link layer protocol: Link Access Procedure Balanced (LAPB) with SDL tools. The textual description of LAPB is fixed in CCITT X.25 standard.

The data-link layer receives the synchronous, serial services of the physical layer and transforms it to the packet level as error-free transport services. The behaviour of the LAPB protocol is nearly completely symmetric at the DTE and DCE side, therefore the description is valid for both sides [DICK-80], [WAG-87].

The elements of the LAPB description are

— processes,
— states,
— enabling conditions,
— signals,
— constants and variables.

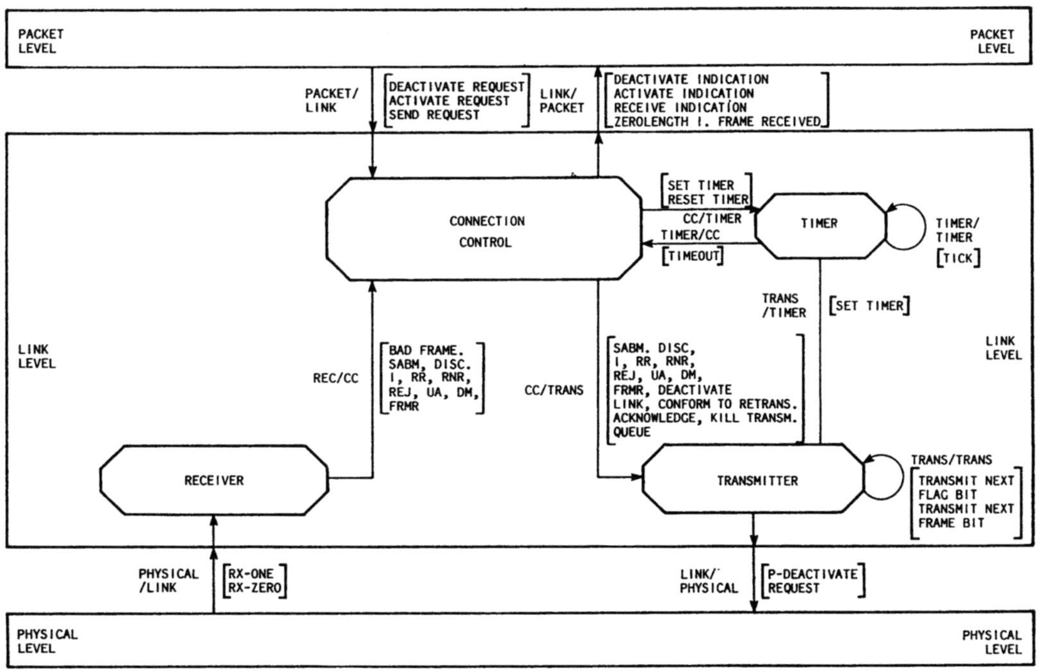

Fig. 9.4. Interactions between the four processes

Four processes create the basic structure of the LAPB description . These are the following:

— Connection Control,
— Transmitter,
— Receiver,
— Timer.

The interaction among the four processes is demonstrated in Fig. 9.4. The four processes are composed of states and programming data (constants and variables). The connections among the states are illustrated as an SDL graph.

The Processes

Now we survey the objectives and possible states of the four processes.

Connection Control Process

Objectives:

— to inform the transmitter which frames should be sent to the partner,
— to receive the frames from the receiver and decide the appropriate response (connection establishment or release, resetting, information transfer),
— to answer the requests of the packet level.

States:

— S1 disconnected,
— S2 set-up requested,
— S3 information transfer,
— S4 disconnected requested,
— S5 frame rejection.

Transmitter Process

Objectives:

— to determine the bit sequence to be transmitted, to insert a flag at the beginning and at the end of the bit sequence,
— to decide about aborting, to send a piggy-backing acknowledgement, to insert a bit,
— to start the timer by frame transmission,
— to detect the deactivation request of the physical layer.

States:

— T1 idle,
— T2 transmitting flag,
— T3 end of flag,
— T4 no priority unnumbered acknowledgement,
— T5 continue,
— T6 transmitting frame,
— T7 discarding I-frame,
— T8 piggy-backing,
— T9 preparing to send priority.

Receiver

Objectives:

— to recognize the flags from the incoming bit sequence, to pick up the frame, to liquidate the bit insertion and control the FCS error detection,
— if a frame is received, to check that the frame belongs to the given LAPB, that its type known, whether the address is correct, and whether the field length is valid,
— if an I-frame is received, then it delivers it to the Connection Control Process.

States:

— R1 looking for flag start,
— R2 looking for flag end,
— R3 looking for frame.

Timer Process

Objective:

— to detect the time-out to the Connection Control Process.

States:

— TM1 reset state,
— TM2 timer waiting state.

Signals

The signals among the processes can be classified into three groups:
— signals between the data-link layer and packet layer,
— signals between the data-link layer and the physical layer,
— signals among the processes.
The signals between the data link layer and packet level:
- deactivate request
  to detect the request of the end of a transmission to the data link layer,
- activate request
  to detect the request of data-link connection establishment,
- send request
  to detect the information sending request to the remote packet level,
  the data content of the signal is the service data unit,
- deactivate indication
  to detect the end of the data-link connection,
- activate indication
  to detect the data connection establishment,
- receive indication
  to detect the arrival of the information from a remote packet level,
  the data content of the signal is the service data unit,
- zero-length I-frame received
  to detect the arrival of the information frame without information field,
  the signals between the data-link layer and physical layer,
- deactivate request
  the request to terminate the data-link layer,
- RX-zero
  to report the reception of the zero bit
- RX-one
  to report the reception of the one bit.

The signals among the processes generally means the signals advancing between two processes:
— between the Receiver and the Connection Control,
— between the Connection Control and the Transmitter Process,
— between the Timer and Connection Control,
— between the Timer and the Transmitter Process.
Some signals realize the feedback to the same process:
— Timer,
— Transmitter Process.
Signals between two processes
- Frames with correct FCS and correct address. The frames are recognized by the Receiver as an implemented command or response.

Frame name	Data content
SABM, DISC	poll bit
UA, DM	final bit
I	poll bit
	N(R)
	N(S)
	information field
RR, RNR, REJ	poll bit/final bit
	N(R)
FRMR	final bit
	information field

- bad frame
  the frame arrived with correct FCS and correct address, but the frame is either unknown or is not implemented in this LAPB realization or the information field is invalid,
- the information necessary for the generation of the frames to be sent.
  The name of the frames is:

  SABM, DISC
  UA, DM
  I
  RR, RNR, REJ
  FRMR

The data-content of the frames is the same as enumerated above. Difference: DM and FRMR has an additional item of information, that sets the timer.

These signals have no address, no FCS and no bit-insertion,

- acknowledge
  to report that a received I-frame must have acknowledged data content: N(R),
- conform-to-retransmit
  to report the arrival of a REJ-frame and the necessity of a retransmission, data content: the sequence number of the first I-frame which should be repeated,
- deactivate link
  request to terminate a data link connection,
- kill trans queue
  to urge the sending of a UA frame,
- reset timer
  to stop the timer clock,
- set timer
  to start the timer clock,
  data-content: the sequence number of the connection control state because of the clock should be started,
- time-out
  to detect the past time.

Past time	Concerning the following state of Connection Control
TIME-OUT 1	S1 Disconnected
TIME-OUT 2	S2 Connection establishment request arrives
TIME-OUT 3A	S3 Information transfer: I-frame
TIME-OUT 3B	S3 Information transfer: S-frame
TIME-OUT 4	S4 Disconnection request arrived
TIME-OUT 5	S5 Frame reject

The Data of the SDL Processes

The interactions among the processes were demonstrated in Fig. 9.4, now we discuss the data belonging to the interactions:

— constants,

— variables.

The constants and their definitions are listed in Table 9.6, the type definition of the constants is in Table 9.7. The variables are enumerated from Table 9.8 to 9.11 grouped according to the processes. The type declaration and the description of variables follow the rules of the PASCAL language.

Table 9.6. Constants of the SDL-graph of the LAPB protocol

CONSTANT	DEFINITION
MANDATORY	
N2=?;	The max. number of retransmitted frames and frames transmitted after the expiry of the timing T1
WINMAX=?;	The max. number of the waiting I-frames following each other to be permitted (K can broaden from 1 to WINMAX)
N1 = ?;	The max. number of bits of an I-frame
T1 - PERIOD=?;	The interval of the timing T1

Table 9.6 (continued)

CONSTANT	DEFINITION
<u>OPTIONAL</u>	
RESETTING	
OPTION # = ?;	O p t i o n 1: the partner can initiate the resetting of the data connection
	O p t i o n 2: reset can be initiated without releasing the data connection
	O p t i o n 3: resetting only after the release of data connection
TIMER	
OPTION # = ?;	O p t i o n 1: the timing starts at the beginning of the frame transmission
	O p t i o n 2: the timing starts at the end of the frame transmission
RNR PROCEDURE	
OPTION # = ?;	O p t i o n 1: transmission or retransmission of the I-frame which has the same sender sequence number as the N(R) of the RNR frame
	O p t i o n 2: sending a supervisory frame which sets the P bit to 1
	O p t i o n 3: nothing happens

Table 9.6 (continued)

CONSTANT	DEFINITION
TIMER RECOVERY OPTION # = ?;	O p t i o n 1: retransmission of the I-frame adjusting the sender state variable to the last received N(R) and setting the P bit to 1 O p t i o n 2: sending the appropriate supervisory command with P bit adjusted to 1

Table 9.7. Type definitions of the constants

```
TYPE                                            type definitions

WINDOW = O..WINMAX ;

STATE = (1, 2, 3A, 3B, 4, 5);

ADDTYPE = (A, B);

FRAMETYPE = (RR, RNR, I, DISC, SABM, DM, UA, FRMR, REJ);

INFOTRANSFERPHASE = (READY,BLOCKED,TIMEOUT-RECOVERY);

INFOSIGNALS = (RR, RNR, REJ, I, TIMEOUT 3A, TIMEOUT 3B,
               SEND REQUEST);

PFTYPE = (POLL, FINAL);

BITSTRING : ARRAY [1..N1] OF O..1 ?

INPUTITEM =

    RECORD

        CASE NAME: INFOSIGNALS OF

            RR, RNR, REJ :

                (NR : WINDOW ;

                CASE PF : PFTYPE OF

                    POLL, FINAL :

                        (PFVAL : O..1));

            I :

                (PFVAL : O..1 ;

                NR, NS : WINDOW) ;

            TIMEOUT3A, TIMEOUT3B, SEND-REQUEST :

END
```

Table 9.8. Variables of the Connection Control Process

CONNECTION CONTROL PROCESS

VARIABLE	INTERPRETATION
RC:O..N2;	number of retransmissions
BV:O..1;	binary value of the poll or final bit of the received signal
V(S):O..WINMAX	sender state variable
V(R):O..WINMAX;	receiver state variable
FINALEXPECTED:BOOLEAN;	true, if we expect a signal with F=1
RESETPROC:BOOLEAN;	true, if the receiver sequence number N(R) of the received signal is valid
BUSY:BOOLEAN;	true, if temporary business exists because I-frames cannot receive
REPORTED:BOOLEAN;	false, if a busy condition starts, but the system did not initiate an RNR response
CLEARED:BOOLEAN;	false, if the busy condition is finished, but the system did not initiate the report of this REJ or RR
DISCARDEDI:BOOLEAN;	true, if the last received I-frame is discarded

Table 9.8 (continued)

VARIABLE	INTERPRETATION
UWE:O..WINMAX;	upper edge of the window
LWE:O..WINMAX;	lower edge of the window
ISENT:BOOLEAN;	DCE(DTE) transmits or repeats an I-frame with N(S) by reception of RNR. If ISENT is true, N(S) is equal with N(R) of RNR. ISENT is false, if N(S) of the I-frame is not equal with the N(R) of RNR
INPUT:INPUTITEM;	it contains the information of the input signal of the state transition as a variable
PHASE: INFOTRANSFERPHASE;	the identification of the information transfer phase
BUFF:ARRAY [O..WINMAX] OF BITSTRING;	the storing of the information field of an I-frame to be sent because of the possible retransmission
SIGNAL:INPUTITEM;	see: "INPUT"
X:O..WINMAX;	if timer Tl has expired, then X should be adjusted to the actual value of the sender state variable
Xl:O..WINMAX;	Xl is the first X value in the timeout sequence
NRMODE:(N, T);	it is equal either with T or with N where T is the time-out NRMODE means the validity check of N(R)

Table 9.9. Variables of the transmitter process

TRANSMITTER PROCESS

VARIABLE	INTERPRETATION
CHANNEL-IS-ACTIVE: BOOLEAN;	true, if the channel is active
CHANNEL-IS-IDLE: BOOLEAN;	true, if the channel is empty
FRAME-SENT: BOOLEAN;	true directly after sending the last bit of the frame
NUM-SENT-ONES: 0..6;	number of 1's in the flag
N(R)-VALUE:0...WINMAX;	value of the possible acknowledgements
Z:0..1;	value of the "SETTIMER" field accompanying the last DM or FRMR signal received by the Transmitter Process

Table. 9.10. Variables of the receiver process

RECEIVER PROCESS

VARIABLE	INTERPRETATION
ONECOUNT:INTEGER;	number of the received and each other following 1-s
ZEROFOUND:BOOLEAN;	true, if a possible 0 arrived in the flag
FSIZE:0..N1;	number of bits in a possible received frame
IDLE-CHANNEL-DETECTED BOOLEAN;	true, if the receiver detected an empty channel
SEQ:ARRAY [0..N1] OF 0..1;	received part of a possible frame
ADDRESS:ADDTYPE;	address of the received frame
TYPE: FRAMETYPE;	name of the received command or response

Table 9.11. Variables of the timer process

TIMER PROCESS

VARIABLE	INTERPRETATION
TIME: 0..TI-PERIOD;	time until the clock is going
ST:STATE;	"state" value of the last rereived signal "SET TIMER"

The Process Diagrams: the SDL-graphs

The SDL-graphs always have a tree-structure therefore it is easy to follow. The S1 *disconnected state* is the root of the tree (Fig. 9.5). S1 belongs to the Connection Control Process. Only one external input event can come in Sl states: the activate request of the packet level.

Internal input events can come either from the Receiver Process (SABM, DISC, DM, I, RR, RNR, REJ) or from the Timer Process (TIME-OUT-1).

The figure demonstrates that the Connection Control changes its state to S3 (information transfer) only on the effect of SABM. The activate request of the packet level transforms the Connection Control Process into the S2 state (connection establishment request arrived). It remains in the S1 state on the effect of the other input events.

The set of poll and final bits can be read from the figure. There are three decisions; the output of two decisions depends on the binary value of the poll and final bit, the third relation is related to the retransmission number, RC.

In the branch *TIME-OUT 1* is a counter incrementing the number of the retransmission by one.

The S1 tree has only one external output: activate indication to the packet level.

Every other output event is internal sending DM to the Transmitter Process and resetting the Timer Process. More fine details can be read from the figure.

The *Receiver Process* looks for the beginning of the flag, the end of the flag and for valid frames. The external input event is either ONE (RX-ONE) or ZERO (RX-ZERO). The operation of the Receiver looking for the beginning of the flag is illustrated in Fig. 9.6. If it is ONE, it is a stuffing bit in the empty channel. If it is zero, then it gets into the R2 state and looks for the end of the flag. If ones arrive, they are counted up to 6. If a zero comes after the sixth one, then the Receiver Process gets to the R3 state and analyses the validity of the frame.

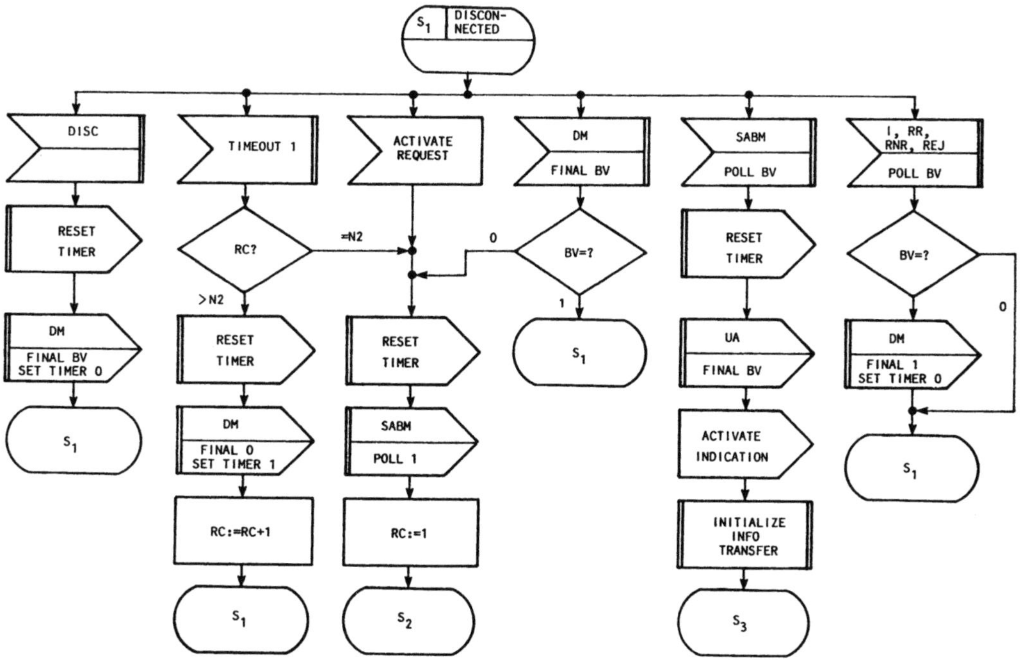

Fig. 9.5. Connection Control, State S1

Every process is described with SDL-graphs. The description is easy-to-follow and its graphical power is a real support in implementation.

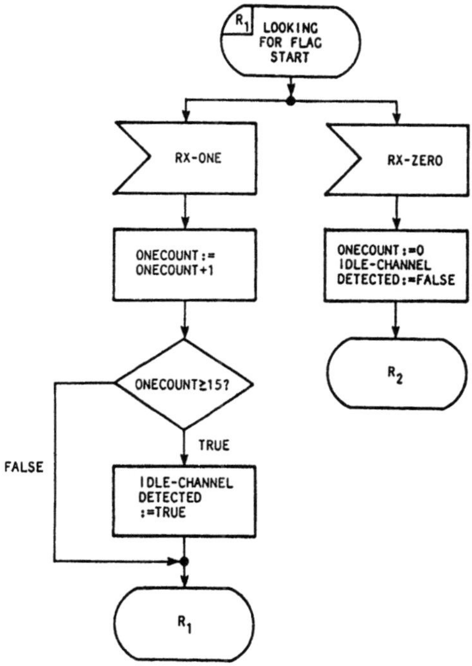

Fig. 9.6. Receiver, State R1

10. FORMAL DESCRIPTION OF TRANSPORT PROTOCOLS

10.1 TRANSPORT PROTOCOL DESCRIPTION WITH FINITE-STATE MACHINE

The testing of state-transitions must be preceded by the unambiguous definition of the state-transitions. The textual protocol specification is not suitable for comparison so it is necessary to describe the standard in closed format which is well-comparable with the implementation. The formal description with finite-state machines [GILL-62] is the most frequently used at present.

One of the most important protocols is the transport protocol. First we introduce the finite-state machine model, then some graphic modules and finally formal descriptions with specification languages.

The M TRANS finite-state machine describes the transport protocol with the following 5-tuple

$$M = (X, Y, S, f_z, f_s)$$

where

- X is the set of input events,
- Y is the set of output events,
- S is the finite set of states,
- f_z is the characteristic function of the outputs,
- f_s is the characteristic function of the states.

The input events of the examined simplified transport protocol (Class 0) are

X = {T_Creq, T_Dreq, T_Cresp, T_DTreq, CR, CC,DT, DR, N_Dind, N_Rind}

The output events of the simplified transport protocol (Class 0) are

Y = {T_Cind, T_Dind, T_Cconf, T_DTind, ERR, N_Dreq}

The states of the simplified, Class 0 transport protocol are

$$S = \{S_1, S_2, S_3, S_4\}$$

where

- S_1 is Idle,
- S_2 is Wait_for_TSAP_Accept_Resp,
- S_3 is Wait_for_TPDV_Connect_Conf,
- S_4 is Data-Transfer.

The state-transitions and outputs can be indirectly determined from the characteristic functions f_z and f_s:

$$z_\nu = f_z\ (x_\nu,)s)$$

$$s_{\nu+1} = f_s\ (x_\nu, s_\nu)$$

where

x_ν, z_ν and s_ν are — in this sequence — the input symbol, the output symbol and state of the M machine at the t_ν moment (ν = 1,2,...).

The characteristic function f_z of a transport protocol represents the fact that an output event y succeeds an input event x when the actual state $s_{\nu+1}$ of the protocol is known. The characteristic function f_z is named the output function.

The characteristic function f_s describes that the next state s_{+1} of the machine will be following the input event x when the actual state of the protocol is s. Therefore f_s is named the next s state function.

The state-transition table of the selected Class 0 transport protocol can be described (Table 10.1) based on the previously introduced terms. Table 10.1 contains the state-transition table of the simplified Class 0 protocol. The header of the columns is the list of input events, the header of the rows is the list of protocol states. An elementary square contains the actual values of the next state/output event. If a hyphen is in an elementary square, it is an unspecified transition. If the hyphen is in the place of the output event, then no output event is generated.

This transport protocol belongs to the category of not fully specified machines because unspecified transitions are present. It is advantageous because it is a strongly connected graph. The M-machine is strongly connected if an input sequence exists transmitting M from any state i into any state j (where i and j can be equal). The demonstrative power of this state-transition table is greater when redrawing it as a state-transition graph (Fig. 10.1). The places of the directed finite graph describe the states of the protocol machine, the arcs represent the state-transitions. The graph has starting and terminating places. The graph arcs are labelled with the input and output events of the protocol machine.

Table 10.1. Transport protocol state table for Class 0

STATE / EVENT	WFNC	WFCC	OPEN	WFTRESP	CLOSED
TCONreq					P3: CR WPCC P4: WFNC P0: TDISind CLOSED P2: NCONreq WFNC
TCONresp				CC OPEN	
TDTreq			[3] DT OPEN		
TDISreq	[1] CLOSED	NDISreq CLOSED	NDISreq CLOSED	DR CLOSED	
NCONconf	CR WFCC				
NRSTind		TDISind [1] [5] CLOSED NRSTresp	TDISind [1] [5] CLOSED NRSTresp	TDISind [1] [5] CLOSED NRSTresp	
NDISind	TDISind CLOSED	TDISind CLOSED	TDISind CLOSED	TDISind CLOSED	
CR					P1: DR ER CLOSED NOT P1: TCONind WFTRESP
DR		TDISind [1] CLOSED	ER OPEN		CLOSED
CC		P8: TCONconf OPEN P6: TDISind NDISreq CLOSED			DR CLOSED
DT			[3] OPEN TDTind		CLOSED
ER		TDISind [1] CLOSED	[2]		CLOSED

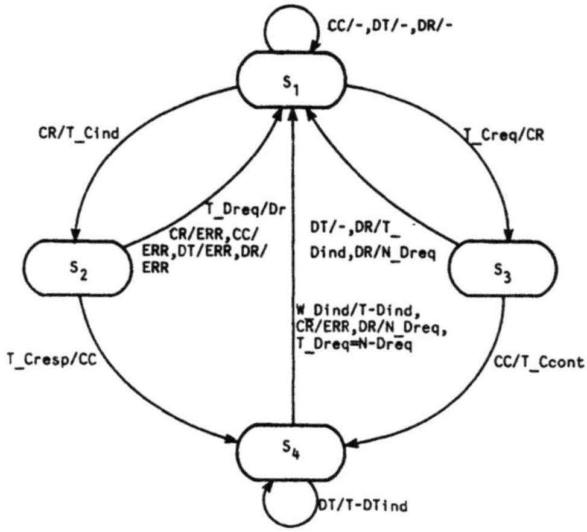

Fig. 10.1. State transition graph of transport protocol Class 0

10.2 TRANSPORT PROTOCOL DESCRIPTION WITH SDL-GRAPH

Different graphical description techniques give a good illustration of the protocol or service specification. The majority of the graphical descriptions are based on extended state-transition models. The term "extension" means that the fairly simple interpretation of the states and transitions is combined with the power of a programming language.

The connection establishment phase of a Class 0 transport protocol is examined with three different graphical descriptions

— state-transition graph,
— extended state-transition graph,
— SDL-graph

and finally we compare the different graphs.

A simple state-transition graph of a Class 0 transport protocol can be seen in Fig. 10.2.

The protocol machine has 4 states

— IDLE,
— WAIT_-FOR_CC,
— WAIT_FOR_-T_ACCEPT,
— DATA_TRANSFER.

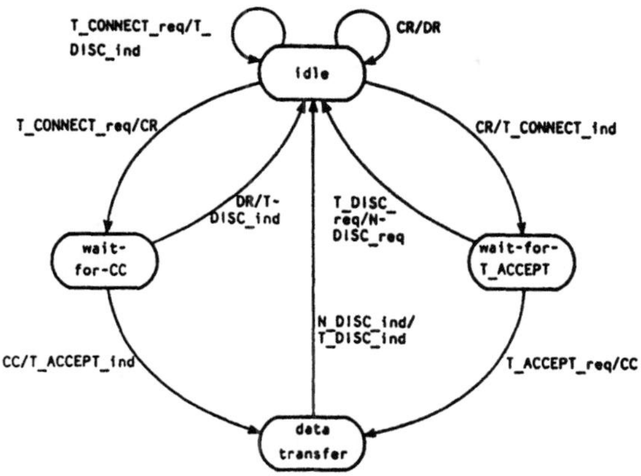

Fig. 10.2. Simple state-transition graph of transport protocol Class 0

The input events are protocol data units and service primitives. The following protocol data units are

— CR Connect Request,
— CC Connect Confirm,
— DR Disconnect Request.

The service primitives are

— T_CONNECT-req,
— T_DISC_req,
— N_DISC_ind.

The output events also are the protocol data units and service primitives. The protocol is symmetric therefore the protocol data units are the same in both directions. The collective names of the service primitives in the opposite direction (CONNECT, DISC, ACCEPT) are identical, only request and indication are transposed.

Extended State-transition Graph of the Transport Protocol

The arcs of the simple state-transition graph are labelled with the names of input event or output event. The output events have a self-contained node character in the extended state-transition graphs placed in a square. The arcs are labelled with the input events and with the enabling conditions (Fig. 10.3).

There are two kinds of conditions in the figure:

+ sign: it can provide service,
- sign: it cannot provide service.

We assigned more input events to a single arc for the simple graph, but only one event for the extended graph. The extended state-transition graph can contain further state variables, in a similar way to the relationships between the transitions and the interaction parameters.

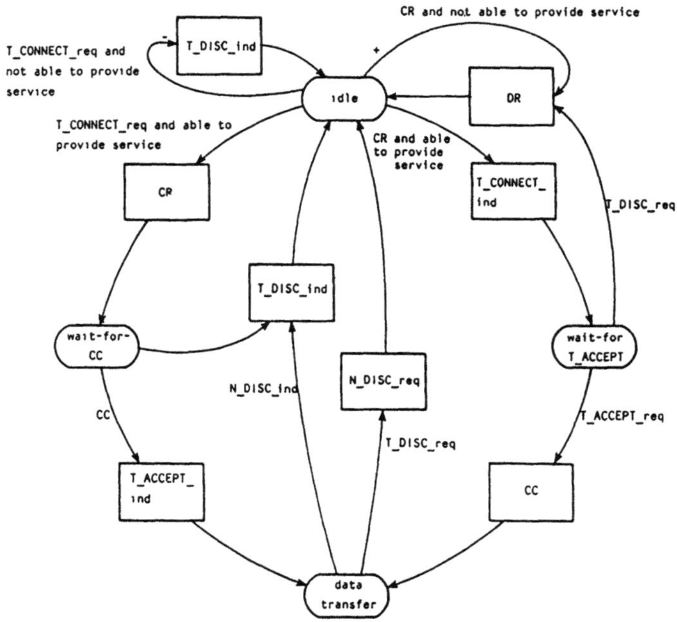

Fig. 10.3. Extended state-transition graph of transport protocol Class 0

SDL/GR Description of the Transport Protocol

The decisions of the SDL graph substitutes the enabling conditions of the extended state-transition graph and the input and output events are drawn in the graph according to the graphic SDL symbols (Fig. 10.4). There are no loops in the figure, all linear paths are easy to follow. The loops — if any — should be opened linearly. The starting point of the figure is the initial state of the protocol, the end of every path is a state. The advantage of the SDL/GR is that it can easily be transformed into the SDL/PR programming language.

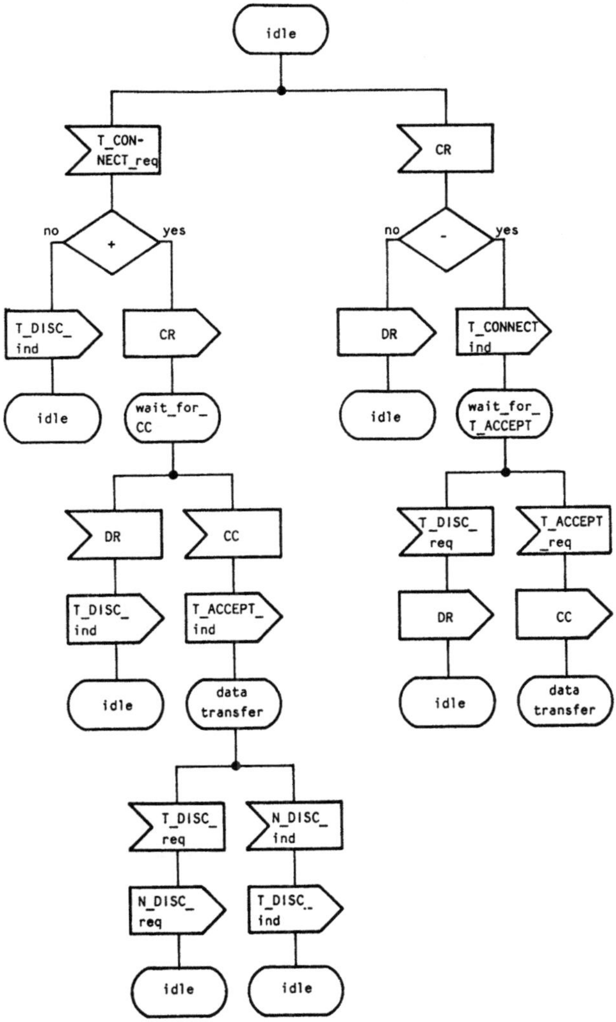

Fig. 10.4. SDL/GR description of transport protocol

10.3 NUMERICAL PETRI NETS OF THE TRANSPORT PROTOCOL

The protocol specification can be decomposed into simple separate Numerical Petri nets [BILL-83]. The transport connection between user A and user B was demonstrated in Section two (see Fig. 2.11). We split the specification into 5 parts for ease of representation and understanding:

— connection establishment (Fig. 10.5),
— normal data transfer (Fig. 10.6),
— TS-user connection release (Fig. 10.7),
— TS-provider connection release (Fig. 10.8),
— queue emptying by the TS-provider (Fig. 10.9).

A list of places and their significance is presented in Table 10.2. The different token types used in the specification are explained in Table 10.3.

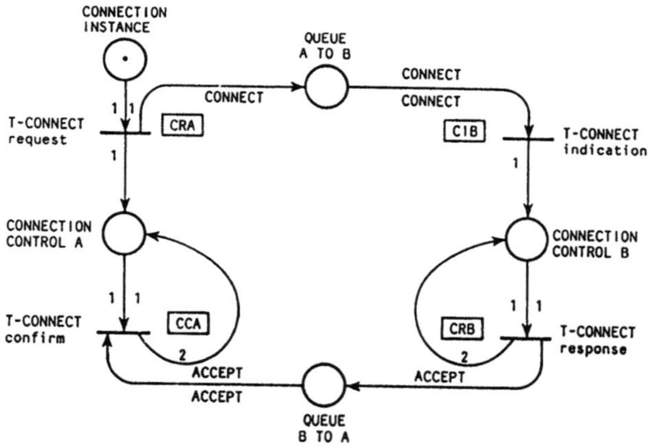

Fig. 10.5. Connection establishment service

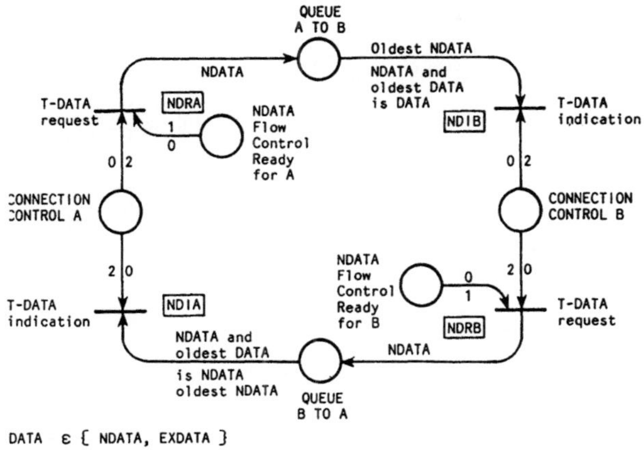

Fig. 10.6. Normal data transfer service

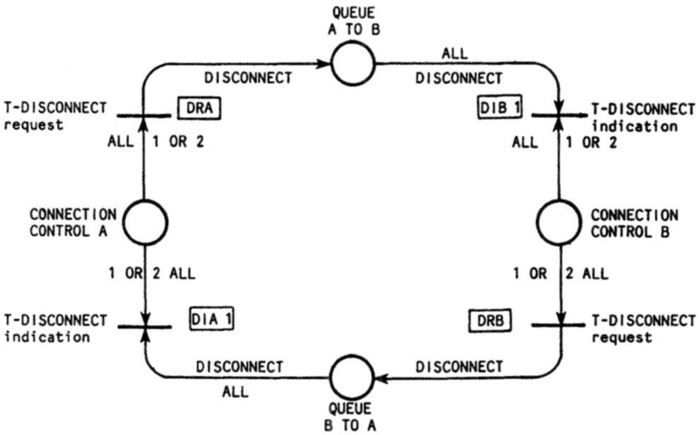

Fig. 10.7. User release

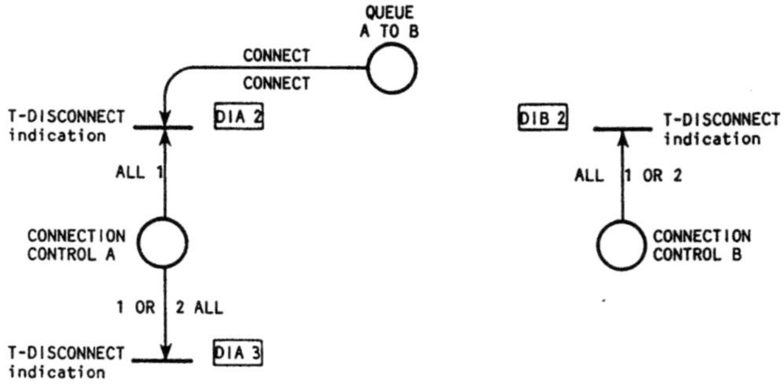

Fig. 10.8. Provider release

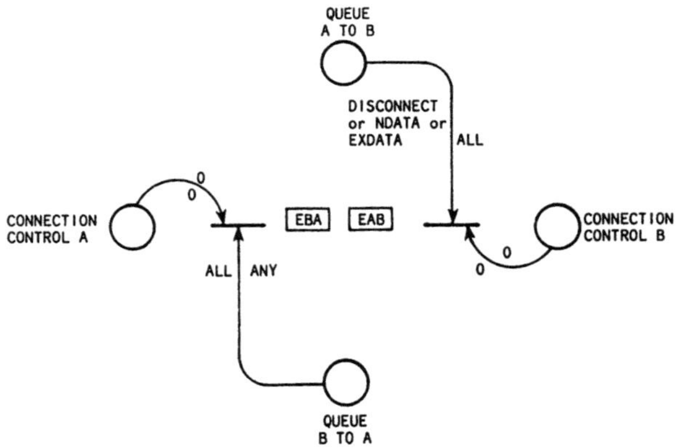

Fig. 10.9. Emptying of queue

Table 10.2. Significance of Places

Place	Significance
CONNECTION INSTANCE	Control place which allows a connection instance to begin.
CONNECTION CONTROL A	Control place which governs the allowable sequences of primitives exchanged between User A and the TS-Provider.
CONNECTION CONTROL B	Control place which governs the allowable sequences of primitives exchanged between User B and the TS-Provider.
QUEUE A to B	Place which models the queue in the TS-Provider for objects flowing from A to B.
QUEUE B to A	Place which models the queue in the TS-Provider for objects flowing from B to A.
NDATA FLOW CONTROL READY FOR A	Control place which allows flow control to be exerted on User A during normal data transfer.
NDATA FLOW CONTROL READY FOR B	Control place which allows flow control to be exerted on User B during normal data transfer.

Table 10.3. Significance of Token Types

Token Type	Significance
Simple	Used for control and represented by a dot.
CONNECT	Represents all parameters contained in the associated T-CONNECT request primitive.
ACCEPT	Represents all parameters contained in the associated T-CONNECT response primitive.
NDATA	Represents all TS User-data contained in the associated T-DATA request primitive.
DISCONNECT	Represents all parameters contained in the associated T-DISCONNECT request primitive.

Transitions

Most of the transitions represent the event of a service primitive exchange between the TS-User and the TS-Provider. These transitions are labelled with the associated service primitive. The labelled transitions occurring on the left side of the NPNs represent service primitive exchanges between User A and the TS-Provider and similarly those occurring on the right side of the NPNs represent service primitive exchanges between User B and the TS-Provider.

The remaining two transitions are not labelled. They allow the TC queues to be emptied when the TC has entered the release phase.

The transitions have been given names which are mnemonics for the events associated with the transitions (e.g. **CRA**:T-CONNECT request from User **A** and **EAB**: Empty Queue **A** to **B**). When the same event is associated with different transitions, the transitions are identified by a numeral following the three letter mnemonic (e.g. DIA3, indicates that this is the third transition associated with a T-DISCONNECT indication being sent to User A). The transition names are enclosed in a rectangle to distinguish them from labels, enabling conditions, and firing rules.

Enabling Conditions for Input Arcs

Enabling conditions for input arcs refer to tokens residing in the associated input place.

Simple tokens are the same as those used in ordinary Petri nets. In NPNs they are not given a name and are represented by dots in places. The number of simple tokens satisfying the enabling condition for an input arc is written next to the arc using the standard drawing convention.

For example, in Fig. 10.6 transition NDRA is enabled if there is a single simple token in the place "NDATA FLOW CONTROL READY FOR A" and there are exactly 2 simple tokens in place "CONNECTION CONTROL A".

A detailed description of Petri Nets can be found in [BILL-83].

As an example, a detailed description of the normal data transfer service is discussed in the next subsection.

Firing Rules

The firing rule for an input arc specifies the tokens which are removed from the associated input place. For simple tokens, the number of tokens to be removed is written next to the input arc according to the standard drawing convention. For example, in the case when transition CRA fires, a simple token is removed from the place "CONNECTION INSTANCE".

For named tokens, the token to be removed is written alongside the input arc.

Conclusions

The Numerical Petri Nets are powerful graphical aids for protocol specification. The trace of all signal flows is easy to follow and different levels of abstraction can be realized.

10.4 DATA FLOW GRAPH FOR TRANSPORT PROTOCOL DESCRIPTION

There are various types of data flow graphs. We use the graphs suggested by Sarikaya [SAR-84]. Every protocol can be decomposed from control flow and data flow. The control flow models the major state changes from one transition to another, the data flow characterizes the parameter changes from input primitive parameters to context variables and from context variables to the output primitive parameters. The input primitives are service primitives and PDUs. The protocol specification contains the parameter list of the input primitives. The parameters of the output primitives are independent and context variables. The independent variables are assigned directly to the output primitives. The context variables are derived from the parameters of the input primitives. The value changes on the context variables are represented as data flow.

The notation (Fig. 10.10) used in data flow graphs is the following:

The input and output primitives as graph nodes are shown in oval shapes containing the list of parameters; the protocol variables and constants are represented as rectangular shapes; circles represent certain functions on data. The input and output primitives, the protocol variables and constants may have an additional node representation, the record. A record is shown in a rectangular shape containing the name of the record and its fields.

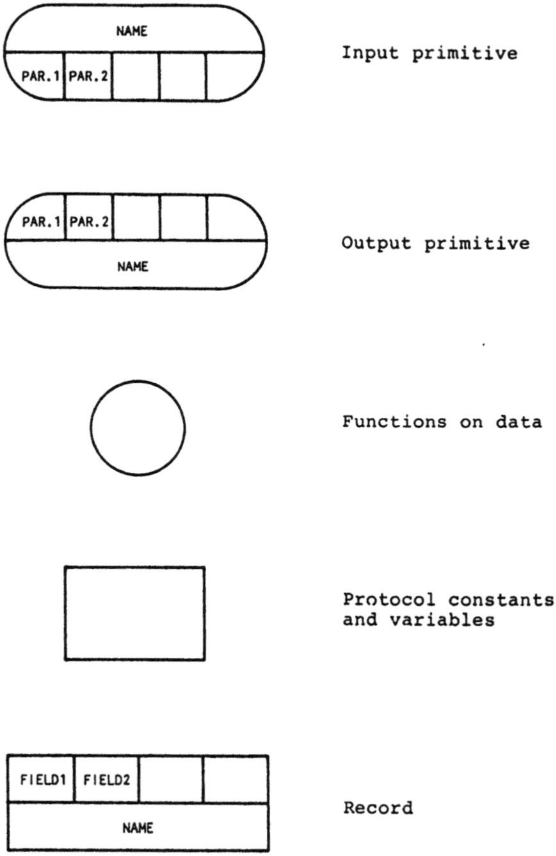

Fig. 10.10. Notation of data flow graph

Many details of the transport protocol can be seen in [SAR-84]. Figure 10.11 shows how the parameters of CR and CC are exchanged, how the source reference becomes a destination reference, and how the TPDU size is determined. The effect of functions can be followed especially during the record construction.

Summarizing, the data flow graph is an effective support with a high visual power enabling one to follow the value conversions between the input and output and to characterize the protocol functions and their usage.

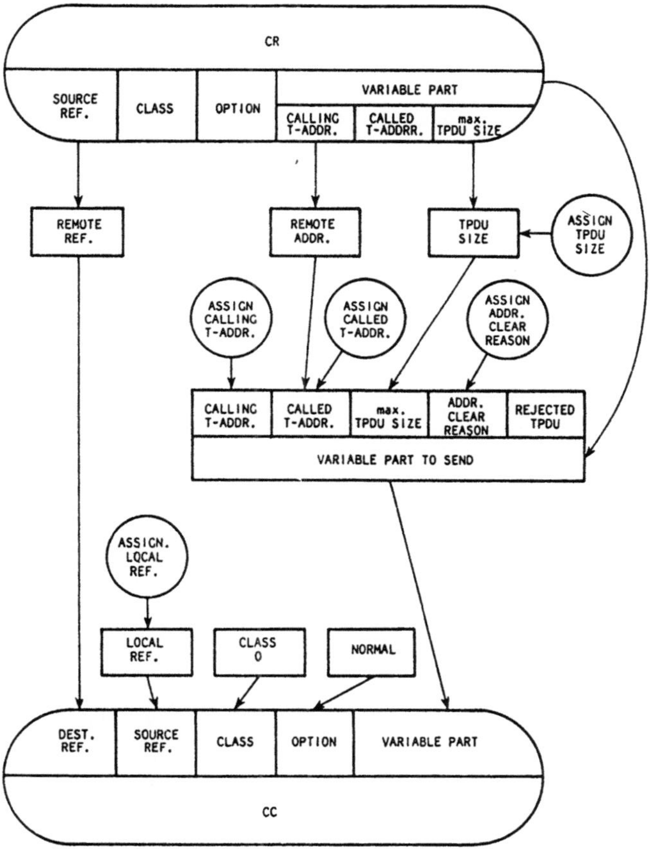

Fig. 10.11. Data flow graph for CR-CC parameter exchange

10.5 LOTOS SPECIFICATION OF THE TRANSPORT PROTOCOL

Structure of the Specification

A protocol can be defined as a set of rules governing the exchange of data between two entities. These rules include the syntax, as data format and the semantics, as control information for coordination and error handling. The LOTOS specification

(Fig. 10.12) can follow the structuring of the rules easily, describing the syntax in the first part and the semantics in the second part of the specification. [BRI-KA-85], [SCO-PAP-86].

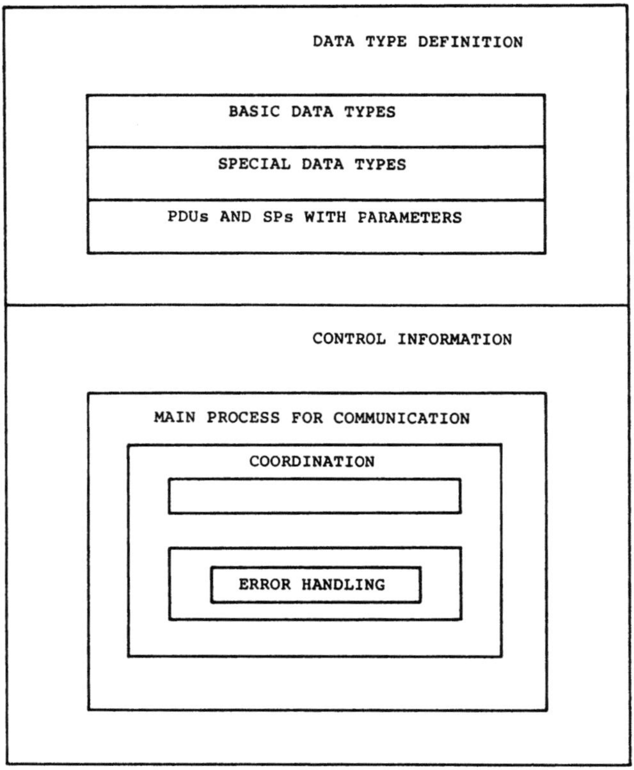

Fig. 10.12. Structure of LOTOS transport protocol specification

The first part deals with the data type definition. The data types to be defined can be classified in three groups:

- the general basic data types (integer, string, Boolean, etc),
- the special basic data types, necessary for the protocol data definition (e.g. address which is used in calling address and called address),
- the protocol data units, service primitives, protocol parameters.

The second part of the specification describes the processes. Essentially, a LOTOS specification is a hierarchy of process definitions. The main process describes the communication between the entities. The main process is composed of other processes. The other processes can either be embedded in each other or can follow each other.

A transport protocol (or any other protocol) can be decomposed of processes in many ways. The following subsections deal with the transport protocol data types and with the processes defining a Transport Connection.

Data Type Definitions

If we want to enumerate the data types which should be defined, we can start from the transport protocol header format. The header format is filled with fixed and variable header primitives. Figure 10.13 shows the header of a CR-TPDU, as an example.

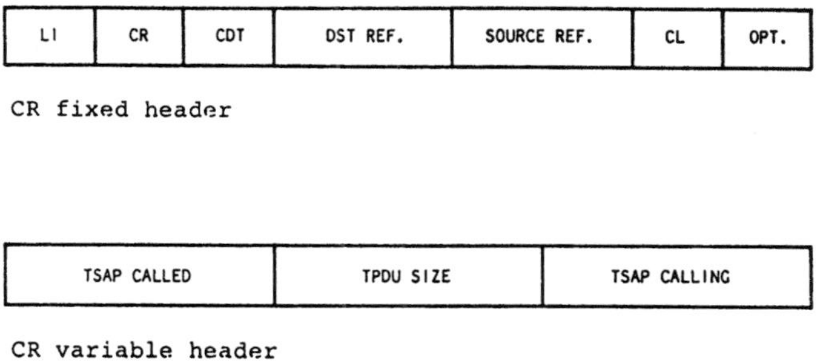

Fig. 10.13. Header of a CR-TPDU

The transport protocol data units are defined in the following way:

```
type TPDU is CODE and REFERENCE and ADDRESS and SIZE and
             QUALITY and EOT and DATA and REASON and
             INV_TPDU and INFO and
  sorts  tpdu
  opns   form_tpdu_from:code reference reference address
             address size quality bool data reason
             inv_tpdu info → tpdu
         codeof:tpdu → code
         sourceof:tpdu → reference
         destinof:tpdu → reference
         cgadr:tpdu → address
         cdadr:tpdu → address
         sizeof:tpdu → size
         qualityof:tpdu → quality
         eotof:tpdu → bool
         usdataof:tpdu → data
         reasonof:tpdu → reason
         inv_tpduof:tpdu → inv_tpdu
         infoof:tpdu → info
  eqns   form_tpdu_from(codeof(t),sourceof(t),destinof(t),
             cgadr(t),cdadr(t),sizeof(t),quality(t),eotof(t),
             usdataof(t),reasonof(t),inv_tpduof(t),infoof(t))=
endtype
```

The TPDU type definition, which refers to the third group of data types, is composed of the header primitives. The header primitives are defined in the second group. As slightly different examples, EOT, CODE and REASON are specified below:

```
type EOT is BOOLEAN and
  opns    EOT_1: → bool
          EOT_0: → bool
  eqns    EOT_1=TRUE
          EOT_0=FALSE


type CODE is
  sorts   code
  opns    CR: → code
          CC: → code
          DR: → code
          DT: → code
          ER: → code
          CONreq: → code
          CONind: → code
          CONresp: → code
          CONcnf: → code
          DISreq: → code
          DISind: → code
          DATAreq: → code
          DATAind: → code
endtype
```

```
type REASON is
  sorts  reason
  opns   reason_in_local_provider: → reason
         reason_in_network: → reason
         reason_in_remote_entity: → reason
         reason_in_user: → reason
         unsuccessful_negotiation: → reason
         overflow_in_reassembling_queue: → reason
         mismatching_reference: → reason
         unknown_address: → reason
         protocol_error: → reason
endtype
```

The first part contains the definition of the following data:

— BOOLEAN,
— NATURALS,
— BIT,
— OCTET,
— STRING,
— QUEUE.

These data types are predefined types in some LOTOS implementations.

This logical chain for data type definition starts with protocol data unit, is followed by header primitives, and ends with general basic data. In the type definition part, all the three parts are enumerated in opposite sequence order, but the logical inference is easier and natural in this way.

The other types of definitions depend on the process-interactions defined by processes which compose the system or protocol.

Specification of the Transport Connection in LOTOS

We use a Transport Connection model consisting of two transport users A and B interacting through the TS Access Point (TSAP). Information is passed between TS users and the service provider through service primitives. The internal structure of the TS provider consists of two queues linking the two transport users, and a queue for each direction of the information flow. Three different phases of behaviour can exist between two TS users: connection establishment, data transfer, and release. The pair of queues is considered to be available for each potential TC; each queue is empty before connection establishment and can be returned by the TS provider to this state at any time. Objects are entered and removed from the queue as a result of interactions at the two TSAPs, subject to the control of the service provider. As we shall see later, these queues are somewhat more complex than FIFO queues. The distinction between service primitives and queue objects suggests a modular structure to describe a single-connection TS consisting of three interacting modules: the TSAP at side A, the TSAP at side B and the pair of queues. This is an example of a modular specification since the end-to-end behaviour of the service provider is described as the result of the interaction between several entities which are, in this case, the queues and the TSAPs.

The process structure [SCO-PAP-86] can be seen in Fig. 10.14. Processes Calling TSAP and Called TSAP specify the local constraints on the user's ability to invoke primitives. Process Transfer specifies the global constraints that relate events taking place at the two SAPs. Process Transfer shows no further static structure, and is parametrized with two queues, one for each direction of communication.

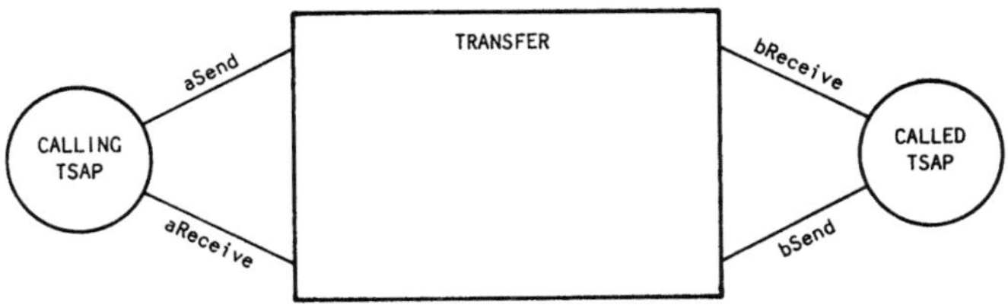

Fig. 10.14. Process structure

The objects that may be placed in the queues are: connect objects (each conveying the parameters of a T-CONNECT primitive), octets of 'normal' data (conveying the parameters of T-DATA primitives, also called TSDUs, for Transport Service Data Units), 'expedited' TSDUs (conveying the parameters of T-EXPEDITED-DATA primitives), and disconnect objects, which can be generated either because of a user's T-DISCONNECT request or because of spontaneous actions by the service provider. The queue has basically a FIFO structure with limited (but not necessarily either fixed or determinable) capacity.

There are some simplifications in the model

— a single connection is described,
— quality of service negotiation is not specified,
— possible loss of message or corruption of data (due to residual errors) are not specified.

The transport Connection Process is in LOTOS

```
process TransportConnection [a,b] (aExp, bExp: Boolean) :=
( CallingTSAP[a, aSend, aReceive](aExp)
|| Transfer[aSend, aReceive, bSend, bReceive](New, New)
|| CalledTSAP[b, bSend, bReceive](bExp)
)\[aSend, aReceive, bSend, bReceive]
endproc (* TransportConnection *)
```

The Transport Connection Process is composed of three processes:
— Calling TSAP Process,
— Called TSAP Process,
— Transfer Process.
The processes have the following specification:

```
process CallingTSAP [User, Send, Receive] (ExpService: Boolean) :=
ConnectReq[User, Send](ExpService)>>
( (ConnectConf[User, Receive]>> def ExpSelected: Boolean in
    DataTransfer[User, Send, Receive](ExpSelected)
  ) [> Disconnect[User, Send, Receive]
)
endproc (* CallingTSAP *)
```

```
process CalledTSAP [User, Send, Receive] (ExpService: Boolean) :=
ConnectInd[User, Receive](ExpService) >> def Terminated, ExpProposed: Boolean in
( [Terminated] → exit
[] [not Terminated] →
  ( (ConnectResp[User, Send](ExpProposed) >> def ExpSelected: Boolean in
       DataTransfer[User, Send, Receive](ExpSelected)
    ) [>  Disconnect[User, Send, Receive]
  )
)
endproc (* CalledTSAP *)
```

```
process Transfer [aSend, aReceive, bSend, bReceive] (qab, qba: TSQueue) :=
      CNX[aSend, bReceive](qab) >> def qab in
         Transfer[aSend, aReceive, bSend, bReceive](qab, qba)
   [] CNX[bSend, aReceive](qba) >> def qba in
         Transfer[aSend, aReceive, bSend, bReceive](qab, qba)
   [] aSend?x:DObject ; Release[bSend, bReceive](Add(x,qab))
   [] bSend?x:DObject ; Release[aSend, aReceive](Add(x,qba))
   [] i ; ( Release[bSend, bReceive](Add(ProviderDisc,qab))
          || Release[aSend, aReceive](Add(ProviderDisc,qba))
          )
endproc (* Transfer *)
```

The detailed description and interpretation of the parameters can be found in [SCO-PAP-86].

11. CONFORMANCE TESTERS

11.1 INITIAL PROTOCOL TESTING

The up-to-date and complex conformance test systems originate from passive and active data communication testing techniques. From passive testers to conformance test centres the way of development leads through the protocol analyzers, from active testers the way is much more direct and better built on the different theoretical methodologies.

In this subsection a short overview is given about the passive aids and protocol analysers, then the initial steps of active tester's development are mentioned. The other subsections discuss the most characteristic conformance test systems.

The first type of passive testers was used as an interface tester realizing the connection to the digital interfaces V.24, V.35, V.36 and X.21/V.11. The tester gives a possibility to disconnect the incoming and outgoing lines of the interface, to switch on the appropriate voltage and to set the logical zero and one. The transmission features of a modem are controlled with a well specified bit-sample. It produces a special text for controlling printers and terminals [TAR-MON-85].

The intelligent data line monitor is a passive observer of the bit sequences crossing the data lines. These monitors are connected between the modem and the terminating equipment. It has many triggering possibilities and therefore can analyse protocol errors, timing problems and the behaviour of the interface. It has a memory and can store the test steps and the results of the analysis.

A protocol analyser also contains an interface tester and a data line monitor, but can simulate the protocol operation. The simulation program is based on state transition tables. The protocol analysers are interactive and either local or remote-controlled.

The first generation of protocol analyzers were used for monitoring and they solved the problems with hardware components. The newer generation has many more software parts for the protocol analysis. Many of these new protocol analysers are based on personal computers with some additional communication interface and application software. They are of two types: one analyses shorter and longer protocol message sequences, the other is developed for the simulation and analysis of the complex protocols of X.25 and SNA networks.

The other trend originated from the application of the active testers. Active testers are connected to the node under test through any transmission medium. The active tester is a program package realized in the test centre and it is suitable for a single-layer-protocol test. The operator should control every instruction manually, but a human decision preceedes the instruction. The centre gives a certification from the test results containing the parameters and description of the test environment and the network characteristics.

An active tester in initial form can solve many tasks, but a further development for advanced testing is useful. One trend is automatic testing without manual interaction. The human decisions are substituted by state-transition table controls which are automatic and continuous. The other trend is the universal, layer independent conformance tester, but the realization of the universality is difficult [AN-81], [BAR-RAY-80], [TAR-ECS-87], [TAR-WAG-87].

11.2 CONFORMANCE TEST CENTRE IN THE NATIONAL BUREAU OF STANDARDS

Requirements

The National Bureau of Standards (NBS) was one of the first organizations to start studying the conformance testing of communication protocols. Since the principles and the procedure of testing developed by NBS have clear logical structure and easy-to-follow temporal ordering, it has been chosen — mainly for didactic reasons — for the present review [LI-NI-83].

It has layered architecture, where every layer has three interfaces:

— upper interface to the service-user,
— lower interface to the service-provider,
— same level interface or the logic interface of the peer-protocol.

The protocol's behaviour should be tested at the three interface levels.

The protocol to be implemented has to comply with the following requirements:

— The protocol has to send correct answers for valid messages received from the service-user and the service-provider interfaces and the peer-protocol.
— The invalid service requests of the interfaces that are not valid have to be rejected.
— Erroneous peer-protocol messages have to be refused.
— The time-outs have to be handled correctly.
— The multiplexed connections have to be handled correctly [LI-CO-83], [LI-NI-83], [LINN-85], [LINN-86].

The Test System

The NBS test system is to be seen in Fig. 11.1. It consists of two parts. The first part is the test centre, the second one consists of the response generator and the response evaluation unit located in the node to be tested. Both in the figure and in the discussion a particular transport protocol's testing is dealt with, but the method is general and it can also be utilized for other layers.

The components of the test system are the following:

— Transport protocol reference implementation;
the protocol to be tested — we assume — is identical with the reference protocol,

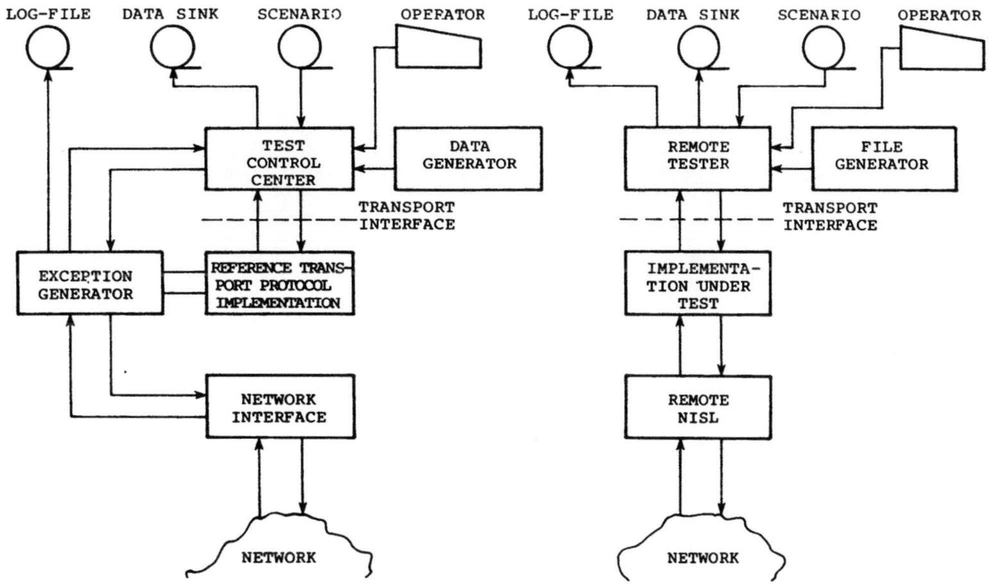

Fig. 11.1. NBS test system

— transport interface,
Network Interface Sub-Layer, (NISL),
this is the upper part of the network layer,
which transforms the abstract service primitives into definite network service primitives,
— remote NISL,
network interface located in the node to be tested,
— transport protocol implementation under test,
— test centre driver,
it controls the test centre and contains the remote tester's process list, so that it allows evaluation depending on test conditions,
— remote tester,
this process cooperates closely with the test centre driver; it generates service-primitives for the protocol under test and receives service primitives from the test centre driver,
— exception generator,
it modifies the messages according to the request of the PDUs of the reference protocol and deletes the received PDUs, when the time-out handling is tested and the recovery procedure is under test. This process is of high importance because the reference protocol can generate and recognize only valid protocol elements and messages.

Operator action starts the tests and the operator can interrupt, if necessary, the running test procedure. All important events are logged in the LOG file. The data generator sets the protocol parameters and the required network environment. The scenario file contains the whole test sequence. The remote scenario is the complement of the test scenario realized in the test centre. By recent implementations the test centre and the remote tester are identical. To keep the integrity of the reference protocol the test control unit also generates transport PDUs and network SDUs.

Test Strategy

There are a lot of service primitives and valid messages and it is not possible to check every alternative. Therefore the most important test patterns should somehow be selected from the practically infinite numbers of possible test alternatives. There are various heuristic methods for the selection of these alternatives.

The scenario is the basic unit of testing. Every test consists of several scenarios. The number of scenarios depends on the required level of quality control. Scenarios consist of commands, the commands consist of service primitives. These commands connect and disconnect links and control the data transport. They provide also certain service primitives for the synchronization between the reference protocol and the protocol under test. The scenarios give valid and invalid combinations of service primitives and, in this way, render possible versatile protocol testing.

Tests are half-duplex. This facilitates both the driving and the response evaluation.

Let us now see a few service primitives applied as commands which are used for transport protocols implementation testing:

CONNECT request	CONNECT indication
CONNECT response	CONNECT confirmation
DATA request	DATA indication
EXPEDITED request	EXPEDITED indication
DISCONNECT request	DISCONNECT indication
CLOSE request	CLOSE indication
UNIDATA request	UNIDATA indication

Complementary service primitives are shown in two columns. The request and the response service primitives contain the commands to the scenario interpreter for the generation of the necessary service primitives for other layers. The indication and the confirmation primitives instruct the scenario interpreter to be ready to receive the appropriate service primitives of the lower layer.

Table 11.1 contains a set of test scenario commands [NI-82].

Table 11.1. The scenario commands

Command	Parameter	Objective
insert	field-value	Insert a parameter
operate	operator, field-value	Perform the given operation (addition, substitution, union or intersection)
output	length, binary string	send the given PDU
remove	-	Remove a parameter
sendup	-	Send the duplicate of the PDU
substitute	field-value	Substitute a parameter value
suppress	-	Suppress (discard) a PDU
swap	length, binary string	Discard a PDU and swap another PDU

11.3 A CONFORMANCE TESTER

Objectives

The importance of protocol conformance testing has increased with the general use of computer networks. In view of this the National Physical Laboratory (NPL), the Gesellschaft für Mathematik und Datenverarbeitung (GMD), and the Agence de l'Informatique (ADI) decided on the common development of test centres. The work is being carried out with financial support from the European Economic Community.

The development has two phases:

1st phase. The realization of the simple physical architecture which is being carried out in the following steps:

— manual conformance tester,

— state-table controlled conformance tester,

— state-table controlled conformance tester with parallel testing tools.

2nd phase. Advanced physical architecture with a network environment manipulator.

Manual Conformance Tester

The operator directly controls the manual system. The service primitives are sent, received and evaluated one after the other. The message sequence and data file help to select and evaluate the primitives — similarly to the scenario file of the NBS system — and make suggestions for the next primitive and reports the right answer arriving on the sent message.

The completed network services between the simple test driver and network protocol coder/decoder helps the test centre to work (Fig. 11.2).

What do the completed services represent?

— the sending of syntactically erroneous messages,

— the replacement of the correct network service primitives into invalid context,

— the detection of the syntactic error of the received message,

— the receiving of the erroneous response [RAY-82], [RAY-HA-81].

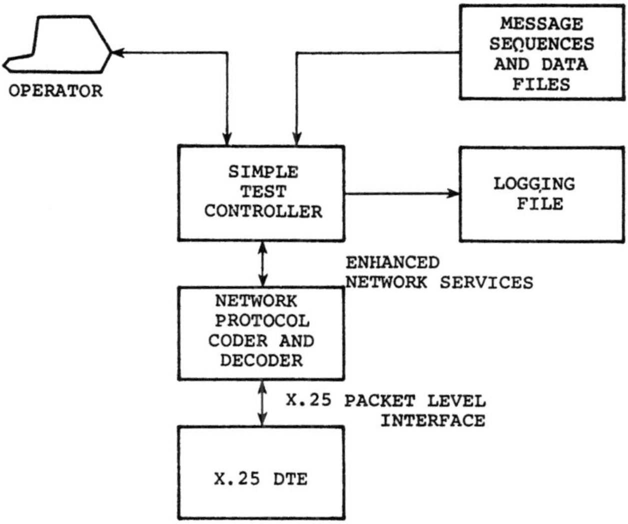

Fig. 11.2. Manual conformance tester

State-transition Table Controlled Conformance Tester

The operator intervene, the testing and has the opportunity to interfere, but essentially the state-transition table generates the test sequence. The state-transition table can be derived automatically from a specification written in a test definition language.

The active tester has the following actions:

— delivery of the messages to the protocol to be tested through the coder/decoder,
— sending messages to the test driver,
— internal actions (to set the timers, to select the parameter limits),
— sending textual messages to the log file.

An active tester can distinguish three kind of events:

— coder/decoder messages,
— test control messages,
— internal events following the internal actions.

The monitoring file evaluates the results after testing and records the traffic data. This system is not suitable for parallel testing.

11.4 TESTING AND DIAGNOSIS AIDS FOR HIGHER LEVEL PROTOCOLS

The GMD has developed a special protocol tester [GIE-83] for testing services of public packet networks, within the TESDI project (TESDI: **TES**ting and **DI**agnosis aid for higher level protocols). The main application field is the testing of TELETEX protocols.

The proposed testing techniques assume that tests are driven under the remote control of the client. The client provides a test responder in order to observe and control the service interface to the (N)-entity under test. The test responder is a possible set of functions which is locally available in the client's system. The client has remote access to the test driver and local access to the test responder and implementation under test.

The main advantage of this approach is that there are no constraints for the specification and implementation of the test responder,

On the other hand, this concept has the disadvantage that the tests are not under the control of the test centre. Such tests require the execution of a commonly agreed test driver-responder-protocol. This must define the predictable and testable reactions of the test responder upon receipt of the PDUs sent by the test driver.

The GMD realized two testing strategies, viz. manually driven tests and automatic tests. The manually driven test has no test driver-responder-protocol, there are no constraints for the test responder and the client can manually control the test driver by test commands. The automatic test is based on test protocols, the test responder is strictly defined and the active test driver reads the test commands from a file.

The test configuration can be seen in Fig. 11.3. The active tester is connected to DATEX-P and can be called as a test partner from the client's system. The test configuration can also be used for functional range testing where a complete test session is started from the terminal using functions such as "echo", "artificial data sources", and "data sinks".

Due to the low transmission rate on the remote operator control connection the trace data are transferred to the central computer. The method of protocol tracing has the advantage that critical steps (even in a long test) can be detected and selectively traced. After the termination of the test the resulting trace data can be read from the central computer and used for post-test analysis.

The test outputs give an easy-to-understand test diagnosis.

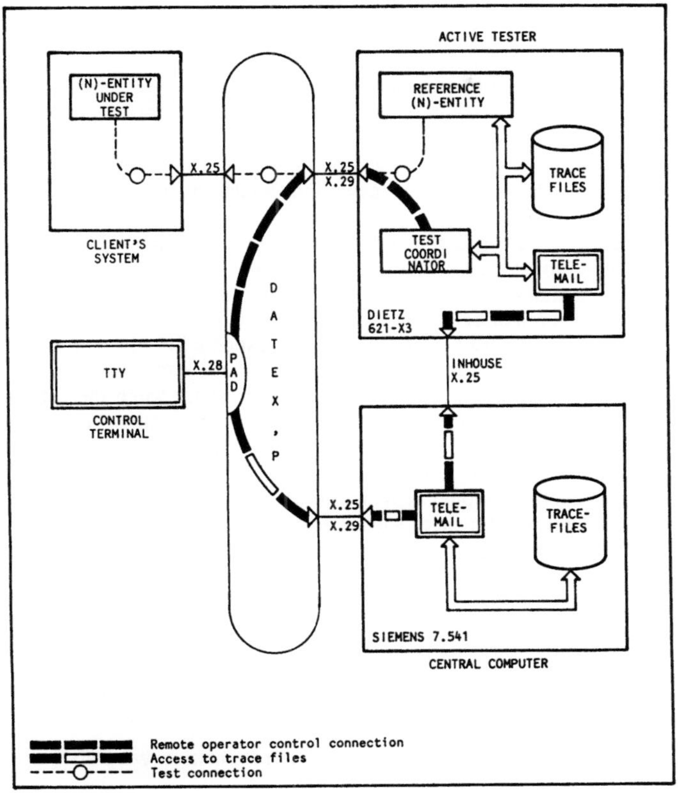

Fig. 11.3. TESDI block diagram

11.5 AT&T PROTOCOL TEST CENTRE

The AT&T (American Telephone and Telegraph Company) developed a test center [SHER-85a, SHER-85b] for packet switched network. The main part is for testing the X.25 communication, especially the DTEs. The testing has three purposes:

- to check the compatibility of DTEs with the X.25 requirements valid for the AT&T packet switched networks,
- to execute some systematic procedures and not a particular testing without consideration of general validity,
- to concentrate on normal cases, but also to recover the errors in abnormal cases.

Requirements

The desired characteristics of a standardized test are:

— Whenever X.25 statements could be interpreted in more than one way, all valid responses should be accepted.

— Tests should be grouped by functions to achieve the following objectives:

- Normal behaviour can be tested independently from error recovery and protocol violations.
- Testing of levels 2 and 3 can be separate as required by the ISO Reference Model (level 2 must be operational before level 3 can be tested).
- Test modules can be rank ordered according to importance so that the tests for normal behaviour can be performed first. For example, it should be possible to examine first the DTE response for all valid frames (packets) in all states then its response to various error conditions.

— There are several states into which the tester cannot force the DTE to enter no matter what stimulating frames or packets it transmits. For example, the tester cannot request the DTE to set up a call without the help of higher level entities. Thus, the DTE should initiate certain actions during the tests and the DTE: manufacturer should provide the necessary means for their initiation,

— Test initialization sequences should accomodate passive, active and hyperactive DTEs. (Whereas passive devices leave the initiative for link level and packet level initializations to the other station, active devices pursue link initialization and a hyperactive X.25 DTE will try to restart the packet level after link level initialization.)

— The tests must verify the correct usage of individual facilities in call set-up packets.

— The tests must verify link level and packet level window rotation and window filling mechanisms.

— The test must verify timer usage and retransmission counters for recovery conditions.

Test Centre Architecture

A basic test centre usually consists of a data base containing all available tests, testing software to execute the tests and microprocessors dedicated to perform X.25 communication and to control the test session. A block diagram representation of the facility is shown in Fig. 11.4. To utilize the capabilities of the testing facility the user dials up the facility to specify test parameters and to schedule a test time. The verification facility then initiates testing at the scheduled date and time through a synchronous dial-up connection. Results are recorded and identified as to whether the tests were successful or not. At the end of a test session, a report is generated with a trace of all tests and all data units exchanged between the facility and the DTE under test. Overall statistics of the session are also included in the report.

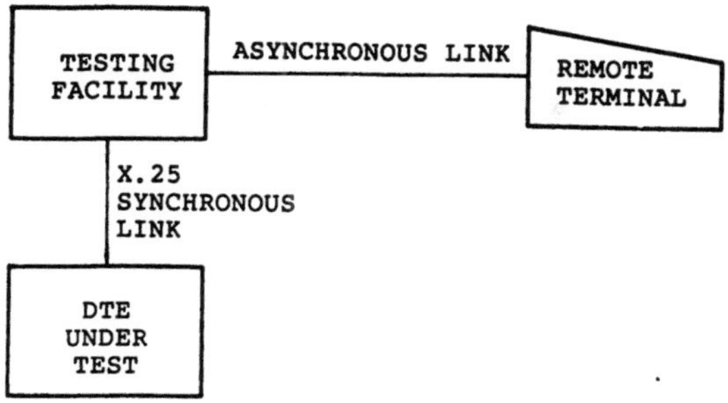

Fig. 11.4. Block diagram representation of testing facility

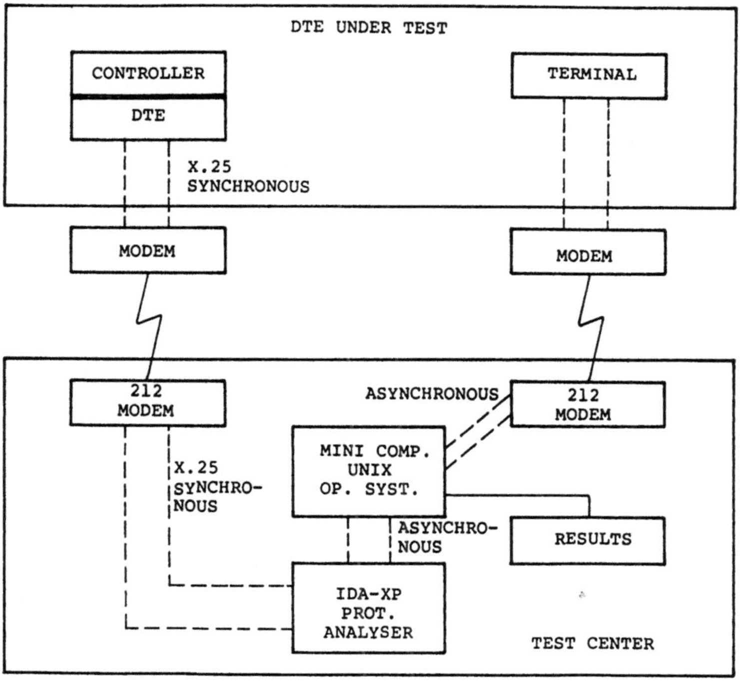

Fig. 11.5. The hardware configuration of the AT&T Test Centre

The facility consists of an IDA-XP X.25 protocol emulator produced by IDA-COM Electronics and an AT&T Technologies minicomputer using the Unix* operating system (Fig. 11.5). The minicomputer is used to download test scripts to the IDA-XP, to record test results and to generate the final report.

As depicted in Figs. 11.4 and 11.5 the IDA-XP has two interfaces: a synchronous interface and an asynchronous one. Communications between the IDA-XP and the DTE under test utilize the synchronous X.25 interface. The synchronous communication path is established via a 1200 bits/sec modem at the test centre and a compatible modem at the DTE test site. Communications between the minicomputer and the IDA-XP are over the asynchronous interface.

The main software components are illustrated in Fig. 11.6. The Session Handler offers an interface between the customer and the test centre. It allows the customer to define in a Parameter File all the necessary protocol parameters and the tests to be run thus minimizing manual intervention. The Test Controller controls the execution of test sequences stored in the Test Sequence Module so that test sequences can be shipped or reordered as needed. The Test Sequence Module consists of the scripts of all test sequences supported by the test centre.

The Protocol Handler has the main role in the test centre. It has three functions:

— to produce formatted frames (packets) and to transmit them to the DTE under test,
— to maintain information on the state of the interface,
— to transmit copies of the frames (packets) sent to and received from the DTE to the Test Logger.

The function of the Test Logger is to record all changes between the Protocol Handler and the DTE for later analysis and printout generation. The Test Analyser determines whether a given exchange of frames (packets) was successful. This determination is based on the information provided by the Test Logger. The Report Generator prints or displays results of the test session based on information stored in the Test Logger. The display may be on remote asynchronous terminals at the testing site and at the test centre.

* Unix is a trademark of AT&T Bell Laboratories.

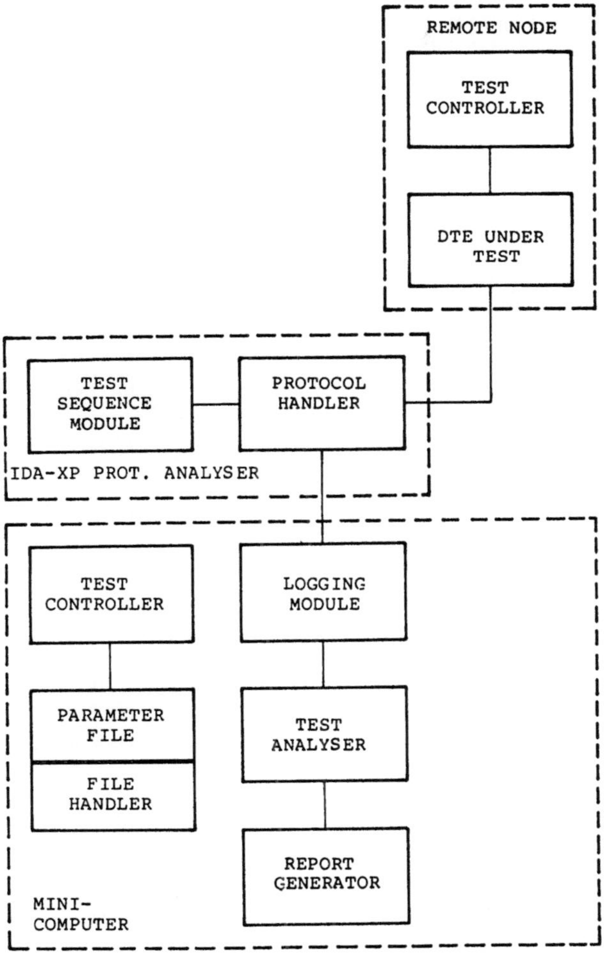

Fig. 11.6. Software configuration of the AT&T Test Centre

Test Sequences

The main test sequences deal with normal frames and packets. Table 11.2 illustrates the testing of the connection establishment at data link level and Table 11.3 describes the testing of a window mechanism (window size = 2) at packet level.

Table 11.2. Testing of the connection establishment

From Tester		From DTE
PASSIVE DTE		
DISC/P=1		
		UA/F=1
		or
		DM/F=1
SABM/P=1		
		UA/F=1
ACTIVE DTE		
DISC/P=1		
		UA/F=1
		or
		DM/F=1
		SABM/P=1
UA/F=1		
	or	
DISC/P=1		
		UA/F=1
		or
		DM/F=1
		DISC/P=1
DM/F=1		
		SABM/P=1
	or	
		DISC/P=1
DM/F=1		
SABM/F=1		UA/F=1

Table 11.3. Test procedure for window size 2 at packet level

From Tester	From DTE
DATA, P(R) = 0	
	DATA, P(S) = 0
RR or DATA, P(R) = 1	
	DATA, P(S) = 1
RR or DATA, P(R) = 2	
	DATA, P(S) = 2
RR or DATA, P(R) = 3	
	DATA, P(S) = 3
RR or DATA, P(R) = 4	
	DATA, P(S) = 4
RR or DATA, P(R) = 5	
	DATA, P(S) = 5
RR or DATA, P(R) = 6	
	DATA, P(S) = 6
RR or DATA, P(R) = 7	
	DATA, P(S) = 7
	DATA, P(S) = 0
	Do nothing!
RR or DATA, P(R) = 1	
	DATA, P(S) = 1

The DTE should also properly handle illegitimate frames and packets and be easily able to recover from errors. There are two categories of illegitimate frames (packets):

— ill-timed frames (packets) which, while being syntactically valid, arrive at an irrelevant or wrong state,

— invalid frames (packets) whose syntax does not conform to the format specifications of X.25.

Further testing is performed at level 3 to check if the facilities are correctly implemented and to examine the various timers and the window rotation mechanism. The testing of time-outs is useful because the various Postal organizations have a free choice of timer values.

Level 3 window mechanism of the DTE is examined, too. The DTE is requested (via a driver) to send a sequence of data packets across the DTE to the testing facility, which acknowledges the first seven packets only. The DTE is expected to refrain from sending the 10th packet (default window size = 2) until the 8th packet is acknowledged. This test and parts of the timer test require the availability of a DTE driver. The level 2 window rotation mechanism is tested in a similar fashion [TAR-87a], [TAR-87b], [TAR-87c].

12. TEST SEQUENCES

12.1 TEST SEQUENCE GENERATION

The control of the semantic conformance of protocols is much more difficult and complicated than the control of syntactic correctness.

It is difficult to test every state-event pair

a) because of the large number of combinations,

b) because of the combinations which never occur under normal circumstances.

The major difficulties are in the enumeration of primitive parameters and in the context (data part).

The maximum that we can do in practice — is to control representative patterns of the state transitions in cases a) and b), then to repeat these tests with the most probably used parameter combinations.

The input message sequence producing the representative pattern of state transitions is called *test sequence*. The active part of the conformance tester sends this test sequence into the node to be tested. The remote test responder generates a complementary test sequence which helps to complete the above enumerated preconditions. The response sequence arriving to the test centre should be compared with the expected responses. That sequence is essentially the second, complementary test sequence.

Next we deal with the generation of the basic test sequence (i.e. the input message sequence required for the basic protocol test) [SAR-BOCH-82].

The systematic test procedures at present are based on the checking methods of finite-state machines. Since the protocols are specified with finite-state machines, therefore we should obviously use the checking methods of finite-state machines [UR-PRO-84].

Transition Tour Method

The transition tour is the most frequently used and simplest method. The input sequence produced by this method is called transition tour [NAI-81].

The preconditions based on the finite-state machine are:

— minimized,

— strongly connected,

— fully defined,

— with fixed initial state.

These preconditions are generally fulfilled, in protocol modelling, besides the "fully defined" feature. If the finite-state machine is not fully defined, then auxiliary states and auxiliary output events should be introduced in order to complete the definition of the protocol.

The steps of the transition tour method are the following:

— either the state-transition graph of the protocol is constructed or the equivalent state-transition table is described,
— starting from the initial state an input event sequence is produced causing the test to pass through each state transition at least once,
— the output event sequence is observed and the executed state transitions are determined from the observed outputs.

We demonstrate the application of the transition tour method with Examples 12.1, 12.2 and 12.3.

Example 12.1

Let us determine the input test sequence of the finite-state machine characterized by the state-transition graph (Fig. 12.1).

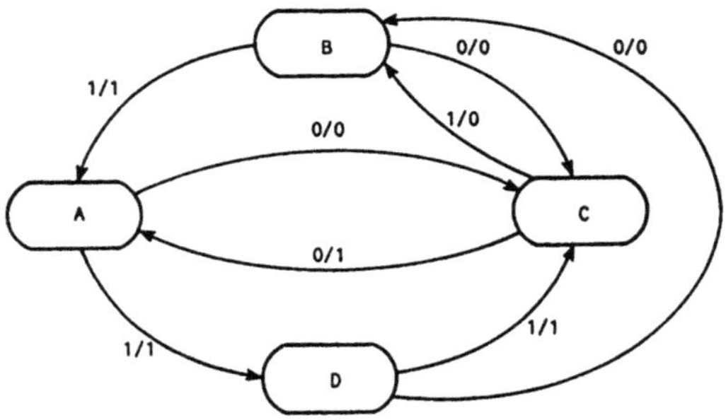

Fig. 12.1. State-transition graph (Example 12.1)

Some transitions are traversed more than once. This example demonstrates a case where the system has a fixed initial state but does not have an unambiguous terminating state.

Example 12.2

A simplified version of the state-transition graph of the Class 0 transport protocol can be seen in Fig. 10.2. The interpretation of the states and events is on page 284. The output event A is an auxiliary variable, introduced in order to get a fully defined protocol description.

Table 12.1 shows how the test sequence is derived from the state-transition graph with the transition tour method.

Table 12.1. Test sequence of a simplified transport protocol (Example 12.2)

Input		CR		T_Dreq		CC		DT		DR		CR	
State	1		2		1		1		1		1		2
Output		T_Cind		DR		A		A		A		T-Cind	
Input		CR		T_Creq		DT		CR		T_Cresp		T-DTreq	
State			1		3		1		2		4		4
Output		ERR		CR		A		T_Cind		CC		N_Dreq	
Input		DT		T_Dreq		CR		CC		CR		DT	
State			4		1		2		1		2		1
Output		T_DTind		N_Dreq		T_Cind		ERR		T_Cind		ERR	
Input		CR		DR		T_Creq		DR		T_Creq		CC	
State			2		1		3		1		3		4
Output		T_Cind		ERR		CR		N_Dreq		CR		T_Cconf	
Input		CR		CR		T_Cresp		DR		CR		T_Cresp	
State			1		2		4		1		2		4
Output		ERR		T_Cind		CC		N_Dreq		T_Cind		CC	
Input		N_Dind		CR		T_Cresp		N_Rind					
State			1		2		4		1				
Output		T_Dind		T_Cind		CC		T_Dind					

Example 12.3

Testing the LAPB

The initial test of this LAPB component is S1 (DISCONNECTED) (see page 250). The test sequence is the following:

Input		ACT.REQ		UA		DISC		DISC		I		ACT.REQ	
State	S1		S2		S3		S1		S1		S1		S2
Output		SABM		ACT.IND		DEACT.IND		DM		DM		SABM	

Input	DISC		SABM		UA		UA		DISC		
State		S1		S3		S2		S3		S1	
Output	DEACT.IND		UA		SABM		ACT.IND		DEACT.IND		

Our test sequence is also

ACT, REQ, UA, DISC, DISC, I, ACT.REQ, DISC, SABM, UA, UA, DISC

12.2 W-METHOD

The next method we discuss is the W-method developed by Tsun Chow for software testing [CHOW-78]. The W-method also is — similarly to the transition tour — a testing strategy based on the finite-state machine theory.

It applies two kinds of testing primitive:

— stimulus,

— response.

We modelize the protocol with a finite-state machine and we produce stimulus-primitives and then we observe the response-primitives.

This method assumes that the following assumptions regarding the finite-state machine should be fulfilled

— it is minimized,

— it is strongly connected,

— it is fully defined,

— it has a fixed starting point.

The assumptions are the same as the assumptions of the transition tour method.

The resultant test sequence is the set of test sequences, more exactly it is the concatenation of P and Z

$$P \cdot Z$$

where

P is the set of input sequences taking the machine from the initial state into each possible state,

Z is the normalized distinguishing sequence.

The set of input sequences, P can best be produced by its test tree. Figure 12.2 demonstrates the test tree determining the set P of the simplified Class 0 transport protocol (page 281). The root of the test tree is the fixed starting point and the initial element of the test sequence.

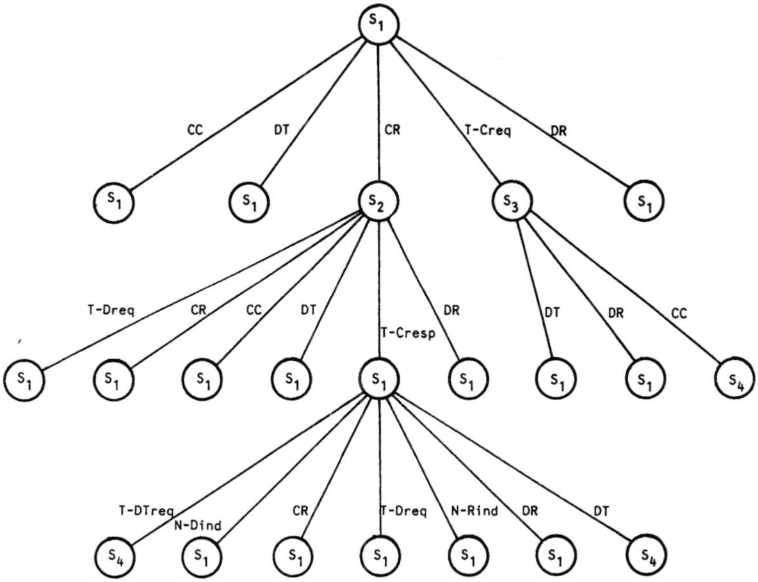

Fig. 12.2. Test tree of transport protocol Class 0

The sequence **Z** is

$$\mathbf{X} \vee \mathbf{W}$$

where

X is the input alphabet, the set of stimuli,

W is the characterizing sequence, distinguishing every state-pair.

It is important to know the input sequence leading into the starting state. In the above example this sequence has only one element: **DR**. The protocol can be in any state, the stimulus **DR** carries it into the initial state S1.

It is important to emphasize that the **W**-method is applied to machines with more states than the specification.

12.3 GÖNENC METHOD

Now we describe a test method worked out by Gönenc for the test sequence generation of sequential circuits [GÖ-70]. The examined sequential machines

— have finite states,
— are synchronized,
— are deterministic,
— are fully specified.

There are further restrictions determining the length of the test sequence:

— there are no equivalent states in the state-transition table,
— the finite-state machine is strongly connected,
— the machine has a distinguishing sequence,
— every possible error occurs during the testing,
— there is no error increasing the number of the states.

Let us review shortly the notations employed in the Gönenc methods. The set of the states is

$$S = \{S_1, S_2, \ldots, S_n\}$$

The set of input symbols is

$$I = \{x_1, x_2 \ldots x_q\}$$

where

n	is the number of states,
q	is the number of input events.

Further notations are

Z	is the output,
X_d	is the distinguishing sequence,
$\delta(S, X)$	is the terminal state function,
$\lambda(S, X)$	is the output sequence function.

If we apply the distinguishing sequence X_d to a machine in state S_j, then we get the following state-string as a result

$$Q_j = \delta(S_j, X_d)$$

and the output-string is

$$Z_j = \lambda(S_j, X_d) \ .$$

A single concrete output is

$$z_{ij} = \lambda(S_j, x_i) \ .$$

The transitional sequence $T(S_i, S_j)$ is the shortest sequence carrying S_i over S_j.

The basic principle of this method is the state recognition. A state is called recognized when it can be identified unambiguously from the output.

A state can be recognized in the following ways:

a) we apply the distinguishing sequence X_d to the state and observe the output. If the response is Z_i on X_d, then the recognized state is S_i. This state is called *d-recognized* and is denoted by subscript d.

b) we apply the homing sequence on the state and observe the output. If the response is Z_i, then the machine will be in the Q_i state. This state is *q-recognized* and is denoted by subscript q.

c) a state can be recognized from a state observed anywhere. Assume that we apply the following input and observe the following output

INPUT	X_d	X	X_d
OUTPUT	Z_i	Z	Z_j

Since this is the q-recognized state, it is certain that the machine was in the $S_k = Q_i$ state before applying the X-sequence.

After the application of X the machine will be in the d-recognized state S_j. In this way we have created the relationp

$$S_j = \delta(S_k, X)$$

If the machine is in state S_k and we apply the input sequence X, then the next state of the machine will be S_j. This state is called t-recognized and is denoted by subscript t.

The parts of the testing sequence are:

— the *initial sequence* carrying the machine to the initial state independently of the starting state at the beginning of the experiment,

— the *α-sequence* recognizing every state and determining every Q,

— the *β-sequence* controlling every transition of the state-transition table.

Since the initial sequence can be determined from the knowledge of the earlier introduced statements (see page 324), therefore we deal only with the generation of the α- and ß-sequence.

α - s e q u e n c e

The input sequence X_d should be applied at least once for every state in order to recognize every state. In this case — assuming the machine operates correctly — we must get minimally n different outputs. There are substrings among the responses given on X_d. These output substrings can be accompanied by n different states.

It is much more reliable to apply the X_dX_d on state S_i of the machine.

INPUT		X_d		X_d	
STATE	S_i		$S_j = Q_i$		$S_k = Q_j$
OUTPUT		Z_i		Z_j	

It is clear that every Q is determined applying X_dX_d on every state.

The generation of *α-sequence*

1st step. If the given state-transition table has more than one distinguishing sequence X_d, then we would have to select the shortest sequence. Draw the diagram X_d and identify the sources.

2nd step. Select a source as the starting state. If there are no sources, then select any state.

3rd step. Apply X_d .

4th step. If the application of X_d leads to an unrecognized state, then we should return to step 3 otherwise we should go forward to step 5.

5th step. Apply X_d.

6th step. If there is an unrecognized source, we should go forward to the 7th step, otherwise to the 8th step.

7th step. Let us apply the T-sequence from the actual state, to the source not yet recognized, then go back to the 3rd step.

8th step. If there is an unrecognized state, then go forward to step 9, otherwise the design of the α-sequence is terminated.

9th step. Apply the T-sequence from the actual state to the state not yet recognized then go back to the 3rd step.

Example 12.4

The machine to be tested is characterized by the state-transition table (Table 12.2). The shortest distinguishing sequence is

$$X_d = 001$$

Table 12.2. State-transition table (Example 12.4)

State \ Input event	0	1
1	2/0	6/0
2	3/0	5/1
3	4/1	2/0
4	5/0	1/0
5	6/0	4/1
6	2/1	5/1

Let us construct the (X_d)-diagram. The (X_d)-diagram can be drawn from the (X_d)-table (Table 12.3). We apply the shortest distinguishing sequence to every state in order to get the (X_d)-table. The related starting and terminating states are connected to each other. The parameter of the arcs is the input sequence/output sequence (Fig. 12.3).

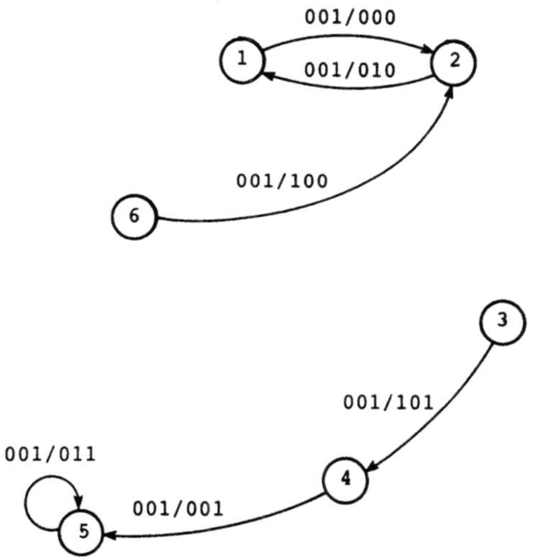

Fig. 12.3. (X_d)-diagram (Example 12.4)

We can see that states 3 and 6 are sources. We can select any of them as the starting state. Let us select state 3, because in this way the resulting α-sequence will be shorter.

The initial sequence is

1100 11100

Table 12.4 shows the steps, input sequences, states, and output sequences.

Table 12.3. (X_d)-diagram (Example 12.4)

Initial state i	Output sequence Z_i	Terminating state Q_i
1	000	2
2	010	1
3	101	4
4	001	5
5	011	5
6	100	2

Table 12.4. States and outputs (Example 12.4)

Sequence number of the applied steps		3		3		3		5		7
Input		001		001		001		001		0
State	3		4^d		5^d		5^q		6^d	
Output		101		001		011		011		0

Sequence number of the applied steps		3		3		3		5		8
Input		001		001		001		001		
State			2^d		1^d		2^d		1^q	
Output		100		010		000		010		

β - s e q u e n c e

The ß-sequence is the state-transition testing sequence, which controls each individual transition of the state-transition table.

The boxes of the state-transition table are called the cells. The ß-sequences controlling each cell are determined from the ß-diagram (Fig. 12.4). Every cell should be tested.

The ß-diagram connects the states, the number of connecting arcs is equal to the number of X_iX_d of the cells. The number of graph places of the β-graph is equal to the number of states of the finite-state machine. The ß-sequence can be read from the ß-diagram.

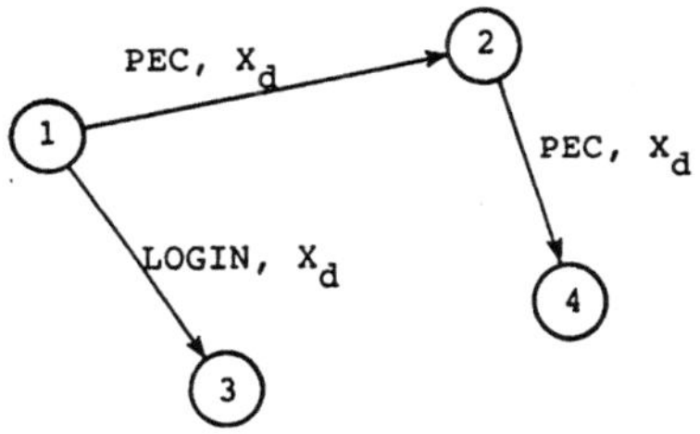

Fig. 12.4. β-diagram

12.4 TESTING THE NATHAN APPLICATION PROTOCOL

The NATHAN application protocol is typically asymmetrical. The primary station (MASTER) is usually a database of the whole network delivering requested data to the secondary stations (SLAVEs). Therefore only the secondary stations can initiate connections (Fig. 12.5). The MASTER protocol is considered as a reference implementation.

The state-transition tables (Table 12.5) and state-transition graphs (Fig. 12.6) characterize unambiguously the NATHAN application protocol. The machine describing the protocol is minimized strongly connected, but not fully defined.

The development of the test sequence is based on the Gönenc method [BRE-MO-85].

The steps of the development are

— the determination of the distinguishing sequence,

— the construction of the α-sequence,

— the creation of the ß-sequence.

The distinguishing sequence can be derived from the successor tree of NATHAN. A typical part of the successor tree is demonstrated in Fig. 12.7. The distinguishing sequence X_d can be read out from the successor tree:

$$X_d = \{PEC, LOGIN, LOGOUT, DATA, ACK\}$$

The NATHAN-responses on X_d are collected in Table 12.6.

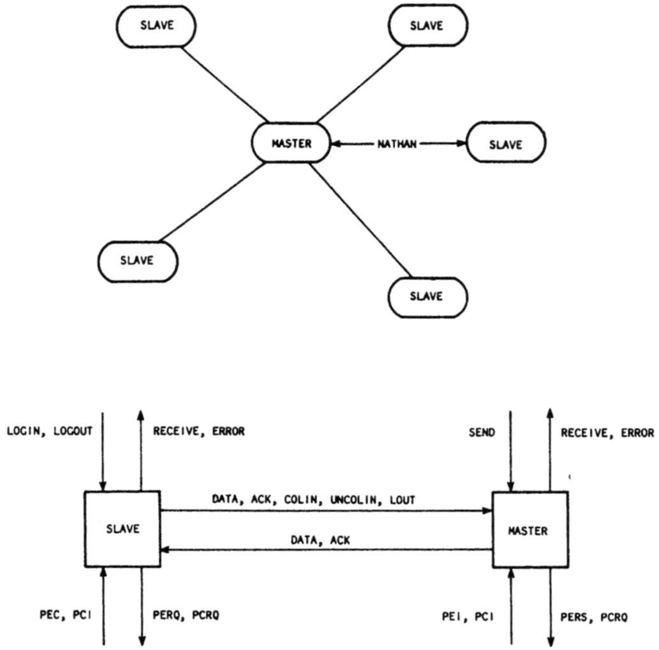

Fig. 12.5. NATHAN application protocol

The initial sequence (IS) is from the state-transition table

$$IS = \{PCI\}$$

The α-sequence is responsible for the tasks:
— to recognize every state,
— to define every "next state".

The X_d-diagram can be constructed by means of the Gönenc method to create the α-sequence (Fig. 12.8) and from the X_d-diagram we can define the α-sequence (Table 12.7).

The ß-sequence deals with the control of every state transition in the state table. The ß-diagram can be seen in Fig. 12.9 using Gönenc's results. The ß-sequence is derivable from the ß-diagram.

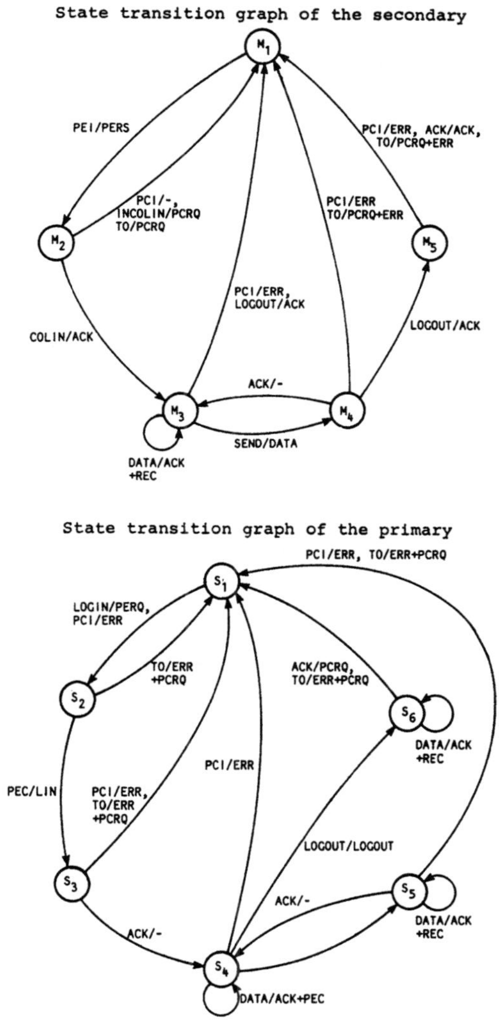

Fig. 12.6. State transition graph of NATHAN

φ(i)-s in the diagram are the input messages belonging to the arc, but the interpretation of the message has no physical meaning. The argument i is the sequence number of the event.

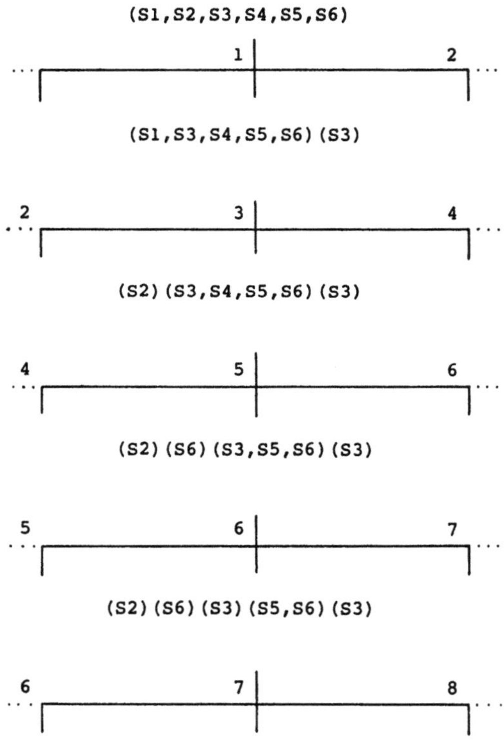

Fig. 12.7. Typical part of the NATHAN successor tree

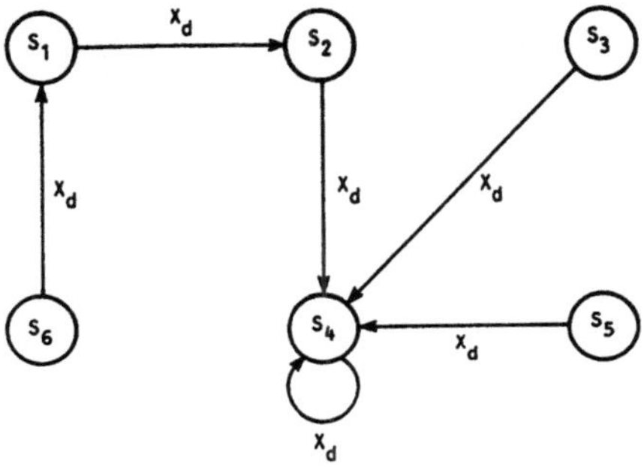

Fig. 12.8. α*-sequence*

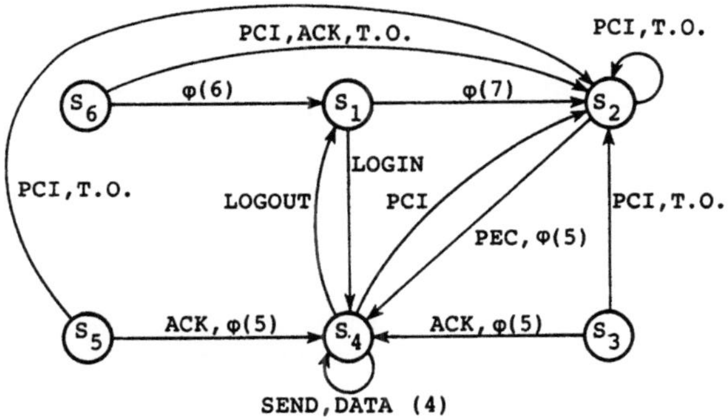

Fig. 12.9. β*-diagram*

Table 12.5/a. NATHAN slave state transition table

EVENT STATE	1 PEC	2 PCI	3 LOGIN	4 SEND	5 LOGOUT	6 DATA	7 ACK	8 T.O.
OFF S1	- 0.0285	- 0.0285	S2/PERQ 0.8 1	- 0.0285	- 0.0285	- 0.0285	- 0.0285	- 0.0285
WAIT FOR CONNECTION S2	S3/LIN 0.08 10	S1/ERR 0.08 1	- 0.008	- 0.008	- 0.008	- 0.008	- 0.008	S1/ERR+PCRQ 0.08 1
WAIT FOR ACK ON LOGIN S3	- 0.008	S1/ERR 0.08 1	- 0.008	- 0.008	- 0.008	- 0.008	S4/- 0.8 10	S1/ERR+PCRQ 0.08 1
DATA TRANSFER S4	- 0.006	S1/ERR 0.06 1	- 0.006	S5/DATA 0.4 20	S6/LOGOUT 0.006 20	S4/ACK+REC 0.4 1	- 0.006	- 0.06
DATA SENT S5	- 0.0058	S1/ERR 0.058 1	- 0.0058	- 0.0058	- 0.0058	S5/ACK+REC 0.058 1	S4/- 0.8 10	S1/ERR+PCRQ 0.058 1
LOGOUT SENT S6	- 0.0058	S1/ERR 0.058 1	- 0.0058	- 0.0058	- 0.0058	S6/ACK+REC 0.058 1	S1/PCRQ 0.8 1	S1/ERR+PCRQ 0.058 1

Table 12.5/b. NATHAN master state transition table

EVENT STATE	1 PEI	2 PCI	3 SEND	4 COLIN	5 INCOLIN	6 DATA	7 ACK	8 LOGOUT	9 T.O.
OFF M1	M2/PERS 0.08 1	- 0.025	- 0.025	- 0.025	- 0.025	- 0.025	- 0.025	- 0.025	- 0.025
WAIT FOR LOGIN M2	- 0.0057	M1/- 0.0057 1	- 0.0057	M3/ACK 0.9 10	M1/PCRQ 0.057 10	- 0.0057	- 0.0057	- 0.0057	M1/PCRQ 0.057 1
DATA TRANSFER M3	- 0.008	M1/ERR 0.08 1	M4/DATA 0.04 20	- 0.008	- 0.008	M3/ACK 0.4 +REC 1	- 0.008	M1/ACK 0.08 20	- 0.008
DATA SENT M4	- 0.0057	M1/ERR 0.057 1	- 0.0057	- 0.0057	- 0.0057	- 0.0057	M3/- 0.8 10	M5/ACK 0.057 10	M1/PCRQ 0.057 +ERR 1
LOGOUT RECEIVED M5	- 0.0077	M1/ERR 0.8 1	- 0.0077	- 0.0077	- 0.0077	- 0.0077	M1/ACK 0.077 1	- 0.0077	M1/PCRQ 0.077 +ERR 1

Table 12.6. Responses of NATHAN slave on X_d

Initial state											Final state
Input		1		3		5		6		7	
State	S1		S1		S2		S2		S2		S2
Output		-		PERQ		-		-		-	
State	S2		S3		S3		S3		S3		S4
Output		LIN		-		-		-		-	
State	S3		S3		S3		S3		S3		S4
Output		-		-		-		-		-	
State	S4		S4		S4		S6		S4		S1
Output		-		-		LOGOUT		ACK		-	
State	S5		S5		S5		S5		S5		S4
Output		-		-		-		ACK+REC		-	
State	S6		S6		S6		S6		S6		S1
Output		-		-		-		ACK+REC		PCRQ	

Table 12.7. α*-sequence*

State	S1	S2	S4
Input	X_d	X_d	K_d
Output	-,PERQ,-,-,-	LIN,-,-,-,-	-,-,LOGOUT,ACK,-
State	S4	S4	S3
Input	X_d	PCI,LOGIN,PEC	X_d
Output	-,-,LOGOUT,ACK,-	ERR,PERQ,LIN	-,-,-,-,-
State	S4	S4	S5
Input	X_d	SEND	X_d
Output	-,-,LOGOUT,ACK,-	DATA	-,-,-,ACK,-
State	S4	S4	S6
Input	X_d	LOGOUT	X_d
Output	-,-,LOGOUT,ACK,-	LOGOUT	-,-,-,ACK,PCRQ
State	S4	S4	
Input	X_d		
Output	-,-,LOGOUT,ACK,-		

12.5 AUTOMATIC TEST SEQUENCE GENERATION

Automatic Testing [FAV-LINN-87]

Automatic testing can substitute for manual testing. There are three test generation methods used most frequently and discussed earlier. These methods are known as:

— the transition tour method,

— the W-method,

— the checking sequence method.

These methods are also used in automatic testing. In this section we focus on partially automated generation of test skeletons for protocols.

These testing methods are valid for classic finite-state machines. The protocols are generally extended finite-state machines with predicates and variables, these abstract machines will be introduced in the next subsection. A set of tools was designed to automatically generate skeletons of test sequences from the ESTELLE specification for a communication protocol. Some tools will briefly be discussed.

Figure 12.10 shows the main idea how the test sequences will be derived from an abstract machine.

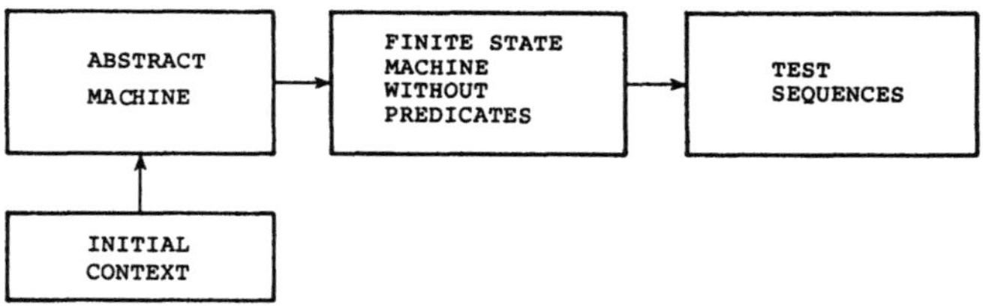

Fig. 12.10. Test sequence derivation from an abstract machine

The Model Used

The test generation method is based on two principles: the separation of the study of an abstract machine by considering only subsets of it, and then the generation of test sequences for each subset.

The abstract machine does not have the same structure as a simple finite state machine. It recognizes input sequences, changes state based upon inputs and predicates, and it has a memory (variables) and it performs actions including emission of outputs during transitions between states, Thus the methods proposed earlier cannot directly be applied to an abstract machine.

The methodology is as follows: decompose the abstract machine into subsets, and then produce test sequences for each of them [CAST-86].

The abstract machine is described as an extended finite-state machine:

$$EFSM = (I,O,S,V,A,F,fs,fo,tr)$$

where

- I is a finite input alphabet,
- O is a finite output alphabet,
- S is a finite set of states,
- V is a finite set of variables,
- A is a finite set of actions on the variables,
- F is a set of Boolean functions on the variables (predicates),
- fs is the "next state"-function,
- fo is the output-function,
- tr is the context transformation function.

The context transformation is described in [FAV-LINN 87]. The context contains the values of the variables at a particular point of time. A possible realization is for an initial context, to consider all the predicates to be true. In this case the abstract machine is reduced to a finite-state machine.

The Tools

The software tools are based on an ESTELLE specification in our example. The ESTELLE protocol specification is structured into modules which are modelled by an extended finite state-machine. The compiler checks to determine if static attributes of the protocol specification are correct, and produces:

— the C-code to be used for the implementation of the procotols,

— a description of the abstract machine modelling the protocol.

Figure 12.11 demonstrates how the test skeleton is derived from the ESTELLE protocol specification. The abstract machine is specified in a formalism which can be used directly by a specific software tool, called LISE [ANS-84]. LISE is a logical and interactive tool for protocol simulation, validation and evaluation.

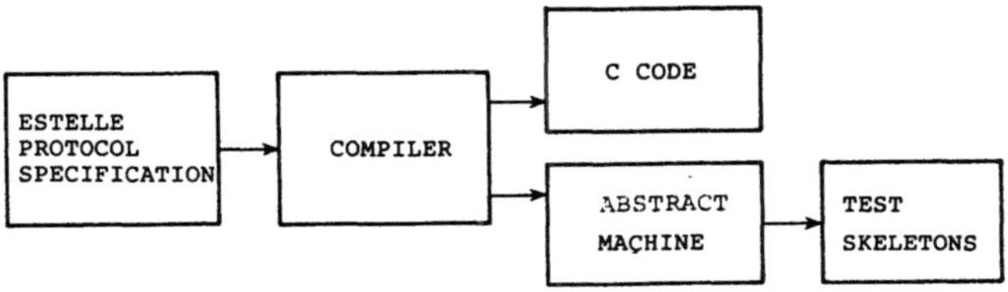

Fig. 12.11. Test skeleton derivation from ESTELLE protocol specification

LISE is composed of two parts:

— VADILOC,

— GAST.

The tool known as VADILOC performs the tasks of verifying some local properties of automata, and validating the communication between two entities using the protocol via a medium with a finite capacity (bounded channels). GAST, the other part, is used to generate test sequences. All the three test sequence generation methods are supported and these are accomplished by the context development method.

These tools have been applied to ISO protocols with promising results.

GLOSSARY

Alternating bit protocol: This is the simplest protocol with sequence numbering O and 1 alternating. It has a single frame type and is suitable for traffic between two stations.

Application entity: A portion of the application process dominant in the cooperation. It is in layer 7 of the OSI model, and it is this layer that determines the interface of the system with the user.

Application layer: Layer 7 of the OSI model. This layer determines the interface of the system with the user.

Channel: A physical or logical path allowing the transmission of information.

Circuit switching: A method of communicating in which a dedicated communications path is established between two devices through one or more intermediate switching nodes. Unlike packet switching, digital data are sent as a continuous stream of bits. Bandwidth is guaranteed, delay is essentially limited to propagation time.

Conformance testing: By this it is tested whether the characteristics of the implemented protocol conform with those given in the protocol specification.

Data circuit-terminating equipment: Equipment, which provides the signal conversion and coding between the data terminal equipment (DTE) and the line. It may be separate equipment or an integral part of the DTE.

Data link layer: Layer 2 of the OSI model. Converts an unreliable transmission channel into reliable one.

Data terminal equipment: Any part of the data station that serves as a data source, data sink or both such as terminals, computers.

Entity: This is a discrete, self contained unit with all its properties defined. It is related to the layers.

Error generator: It sends and receives sequences of messages being erroneous syntactically, semantically, or temporally.

Event: The atomic form of an interaction in LOTOS.

Flag: In communications a bit pattern (of six consecutive 1 bits) used in many bit-oriented protocols to mark the beginning and the end of a frame.

Flow control: The function performed by a receiving entity to limit the amount or rate of data that is sent by a transmitting entity.

Formal description: An unambiguous way of describing a (protocol) specification concisely. The abstraction level of the formal description extends from the network architecture to the dialogue realized by a protocol. It needs to be easy-to-handle and should have general validity.

Formal language: A set of sentences which are finite strings drawn from a finite vocabulary.

Frame: A bit sequence sent serially over a communication channel. It forms the block of data to be transmitted.

Full-duplex transmission: Data transmission in both directions at the same time.

Half-duplex transmission: Data transmission in one but either direction at a time.

HDLC: A very common bit-oriented data link protocol (OSI layer 2).

IUT: (Implementation Under Test). The part of a real open system which is to be studied by testing.

Layer: The functional block of the Reference Model. There are 7 seven layers in the OSI model; these layers identify the functions of the layers and the relationships between them.

Modem: Modulator/demodulator. Transforms a digital bit stream into an analog signal (modulator) and vice versa (demodulator).

Multiplexing: The combining of multiple data channels into a single transmission medium.

Multipoint: A configuration in which more than two stations share a transmission path.

Network layer: Layer 3 of the OSI reference model responsible for routing data through a communication network.

Node: The point of a network or graph through which an arc or trajectory goes in or out from the nodes.

OSI: (Open Systems Interconnection). This is a logical structure for network operations standardized by the ISO. Computers or devices that are OSI-compliant are able to communicate with each other. (See Reference Model.)

Packet switching: A method of transmitting messages through a communication network: long messages are subdivided into short packets, and they are then transmitted. It is done by a packet store-and-forward system and no dedicated path is established, each packet contains a destination address and is passed from source to destination through intermediate nodes.

Performance measurement: It determines how fast and how reliable information transfer can be provided by a given protocol in operation for different loads.

Physical layer: Layer 1 of the OSI reference model, it is concerned with the electrical, mechanical and timing aspects of signal transmission over a medium.

Point-to-point: A configuration in which two stations share a transmission path.

Presentation layer: Layer 6 of the OSI reference model, it is concerned with data format and display.

Process: Generally a software application: it is a sequence of operations that produces a specified result.

Protocol: The fundamental concept of the Reference Model, it defines the syntactic, semantic and temporal rules of cooperation between two entities.

Reference model: It is an abstract model in which every real open system is described by an abstract open system as a whole without preconditions for its internal organization. It identifies the main functional blocks in seven layers.

Reference protocol: A faultless, correct implementation of a protocol, that conforms with the specification. During conformance testing the IUT is compared with the reference protocol.

Routing: The process of selecting the correct circuit path for a message.

Session layer: Layer 5 of the OSI model. It is this layer that arranges a logical connection (session) between two communicating processes or applications.

Service: This is the capability of a layer which is provided to the entities of the upper layer at the boundary between the two layers.

Service access point: The logical interface between the (N)- and (N+1)-entity.

Specification: Describes the protocol exactly with formal description modules.

Teletex: A text communications service which provides message preparation and transmission facilities.

Test driver: An element of the conformance tester, it allows the testing to start and makes observations and evaluations.

Test responder: Abstraction of the means of providing — during test execution — control and observation of the upper service boundary of the IUT.

Test sequence: A tool for examining the protocol, it is built of protocol data units and service primitives.

Test sequence generation: The process which results in an executable test sequence, it has several methods mostly based on FSM representation of the protocol.

Timeout: A protocol defines the temporal rules of the communication. Protocol entities have timers, and when a timer expires it means a timeout event.

Transport layer: Layer 4 of the OSI model. It provides reliable, transparent transfer of data between endpoints.

Verification: Checking the logical correctness and proper operation of the protocol.

V.24 (CCITT): List of definitions for interchange circuits between data-terminal equipment and data-circuit terminating equipment.

Window: A flow-control mechanism in data communications. The window size is equal to the number of frames or messages that can be sent from a transmitter before any acknowledgement is required from the receiver.

X.21 (CCITT): A network access standard for connecting stations to a circuit-switched network. Includes OSI layer 1-3 functionality.

X.25 (CCITT): A network access standard for connecting stations to a packet-switched network. Includes OSI layer 1-3 functionality.

X.409 (CCITT): A standard developed for a Message Handling System. Includes OSI layer 6.

Z.100-Z.104 (CCITT): A set of recommendations which give the general definition, the basic concepts,the structuring concepts, the refinement, and the data definition of SDL.

REFERENCES

[AGE-79] Agerwala T.: "Putting Petri Nets to Work", Computer, Vol. 12, No. 12, December 1979, pp. 85-94.

[AHO-SAR-87a] Ahooja R., Sarikaya B.: "An Unified Data Flow Model for ESTELLE and LOTOS Specification", Report, Concordia University, Montreal, 1987.

[AHO-SAR-87b] Ahooja R., Sarikaya B.: "Comparing Normal Forms Obtained from ESTELLE and LOTOS Specifications", in [SAR-BOCH-87], pp. 373-384.

[AND-83] Andrásfai B.: The Theory of Graphs, Flows and Matrices. Akadémiai Kiadó, Budapest, 1983 (in Hungarian).

[AN-81] Ansart J.P.: "Development and Certification of Standardized Protocols", in [CSA-81], pp. 477-487.

[AN-87] Ansart J.P. et al.: "Software Tools for ESTELLE", in [SAR-BOCH-87], pp. 55-62

[AN-CHA-DEM-86] Ansart J.P., Chari V., Dembinski P.: "Deriving Implementation Oriented Specifications within ESTELLE", Sedos Project ESPRIT Report SEDOS/084.

[ANS-84] Ansart J.P., Castanet R., Guitton P., Rafiq O.: "Some Operational Tools in an OSI Protocols Study Environment", Computer Communication Review, Vol. 14, No. 2, 1984, pp. 156-161.

[AZ-AY-BE-78] Azema P., Ayache J.M., Berthomieu B.: "Design and Verification of Communication Procedures: A Bottom-Up Approach", Proc. of the Third International Conference on Software Engineering, 1978, or in [SUN-81], pp. 135-141.

[BAR-84] Barchanski A.: "Relationship between the Errors and Layered Structure" Personal communication, 1984.

[BAR-SCA-69] Bartlett K.A., Scantlebury K.A., Wilkinson P.T.: "A Note on Reliable Full-Duplex Transmission over Half-Duplex Link", Commun. ACM, Vol. 12, No.5, 1969, pp. 260-261.

[BAR-RAY-80] Bartlett K., Rayner D.: "The Certification of Data Communication Protocols", NBS Trends and Applications Conference, May 29, 1980, pp. 12-17, or in [SUN-81], pp. 383-388.

[BEL-SARA-86] Belina F., Saracco R.: "CCITT-SDL: Overview of the Standard and its Application in the Area of Protocol Specification", in [CSA-86], pp. 505-518.

[BER-TER-81] Berthelot G., Terrat R.: "Petri Nets Theory for the Correctness of Protocols", in [SUN-81], pp. 325-342.

[BER-80] Bertine H.V.: "Physical Level Protocols", IEEE Trans. on Comm. Vol. COM-28, April 1980, pp. 433-444.

[BILL-83] Billington J.: "Abstract Specification of the ISO Transport Service Definition using Labelled Numerical Petri Nets", in [RUD-WE-83], pp. 173-185.

[BJOR-70] Bjorner D.: "Finite State Automaton - Definition of Data Communication Line Control Procedures", AFIPS Conference Proceeding, Vol. 37, 1970, pp. 477-491.

[BOCH-76] Bochmann G.v.: "Finite State Description of Communication Protocols", Technical Report No. 236, Univ. Montreal, Dept. of Inform., 1976.

[BOCH-78] Bochmann G.v.: "Finite State Description of Communication Protocols", Computer Networks, Vol. 2, Oct. 1978. pp. 362-372.

[BOCH-80] Bochmann G.v.: "A General Transition Model for Protocols and Communication Services", IEEE Trans. on Comm. Vol. COM-28, No. 4, April 1980, pp. 643-650.

[BOCH-87] Bochmann G.v.: "Usage of Protocol Development Tools: The Results of a Survey", in [RUD-WE 87], 1987.

[BOCH-GE-77] Bochmann G.v., Gecsei J.: "A Unified Method for the Specification and Verification of Protocols", Information Processing 77, B.Gilchrist (ed.), IFIP, North-Holland Publ. Co., Amsterdam, 1977.

[BOCH-SUN-80] Bochmann G.v., Sunshine C.: "Formal Methods in Communication Protocol Design", IEEE Trans. on Comm. Vol. COM-28, April 1980. pp. 624-631.

[BOH-84] Bohus M.: "Simulation Research of Computer Network Protocols", M. Sc. Thesis, Szeged, 1984 (in Hungarian).

[BOL-DENI-86] Bolognesi T., de Nicola R.: "A Tutorial on LOTOS", CNUCE/CNP, Pisa, Italy, March 1986.

[BRE-MO-85] Bremer J., Mondvai G., Tarnay K., Tibor J.: "Some Experiences with Test Sequence Generation in Application Layer", in [YEM-85], pp. 623-636.

[BRI-86] Brinksma E.: "A Tutorial on LOTOS", in [DIAZ-86], pp. 171-194.

[BRI-FE-LO-87] Briand J.P., Fehri M.C., Logrippo L., Obaid A.: "Executing LOTOS Specifications", in [SAR-BOCH-87], pp. 73-84.

[BRI-KA-85] Brinksma E., Karjoth G.: "A Specification of the OSI Transport Service in LOTOS", in [YEM-85], pp. 227-252.

[BRI-SCO-STE-87] Brinksma E., Scollo G., Steenbergen C.: "LOTOS Specifications, their Implementations and their Tests", in [SAR-BOCH-87], pp. 349-360.

[BU-SA-69] Busacker R.G., Saaty T.L.: "Finite Graphs and Networks. An Introduction with Applications." McGraw Hill Book Company, New York, 1969.

[BU-82] Bux W., Kümmerle K., Truong H.L.: "Data Link-control Performance: Results Comparing HDLC Operational Modes", Computer Networks, 1982, pp. 37-51.

[CAR-86] Carchiolo V., Faro A., Mirabella O., Pappalardo G., Scollo G.: "A LOTOS Specification of the PROWAY Highway Service", IEEE Trans. on Computers, Vol. C-35, No. 11, Nov. 1986.

[CAST-GU-85] Castanet R., Guilton P., Rafiq O.: "An Automatic System for Study of Protocols: a Presentation and Critique Based on a Worked Example", in [YEM-85], pp. 111-126.

[CAST-86] Castanet R., Sijelmassi R.: "Local Tests for Distributed Testing", in [CSA-86], pp. 663-676.

[CERF-72] Cerf V.G.: "Multiprocessors, Semaphores and a Graph Model of Computation", Ph. D. Thesis, UCL2-ENG-7223, 1972.

[CHOM-75] Chomsky N.: The Logical Structure of Linguistic Theory, Plenum, New York, 1975.

[CHOW-78] Chow T.S.: "Testing Software Design Modeled by Finite-state Machines", IEEE Trans. on Software Engineering, Vol. SE-4, No.3, May 1978, pp. 178-187.

[CSA-81] Networks from the user's point of view. Csaba L., Szentiványi T., Tarnay K. (eds.), North-Holland Publ. Co., Amsterdam, 1981.

[CSA-86] Computer Network Usage: Recent Experiences. Csaba L., Tarnay K., Szentiványi T. (eds.), North-Holland Publ. Co., Amsterdam, 1986.

[COU-PE-AY-85] Courtiat J.P., Pedroza A., Ayache J.M.: "A Simulation Environment for Protocol Specifications Described in ESTELLE", in [DIAZ-86], pp. 297-312.

[DAN-BRE-75] Danthine A., Bremer J.: "Communication Protocols in a Network Context", ACM SIGCOMM - SIGOPS Interface Workshop on Interprocess Communications, Santa Monica California, 1975.

[DAN-BRE-76] Danthine A., Bremer J.: "An Axiomatic Description of the Transport Protocol of CYCLADES", Professional Conference on Computer Networks and Teleprocessing, Aachen, 1976.

[DAN-77] Danthine A.: "Petri Nets for Protocol Modelling and Verification", COMNET '77 Budapest, Working Papers, Vol. II, pp. 663-685.

[DAN-80] Danthine A.: "Protocol Representation with Finite-State Models", IEEE Trans. on Communications. Vol. COM-28, No. 4, April 1980, pp. 632-643.

[DAU-BAR-79] Davies D.W., Barber D.L.A., Price W.L., Solomonides C.M.: Computer Networks an Their Protocols. Chichester-New York etc. Wiley, 1979.

[DEN-8O] Dennis J.B.: "Data Flow Supercomputer", IEEE Computer, Nov. 1980.

[DIAZ-86] "Protocol Specification, Testing, and Verification", Diaz W.M. (ed.). North-Holland Publ. Co., Amsterdam, 1986.

[DIAZ-82] Diaz M.: "Modeling and Analysis of Communication and Cooperation Protocols using Petri net based models", Computer Networks, 6, 1982. pp. 419-441.

[DICK-80] Dickson G.J.: "Formal Specification Technique for Data Communication Protocol X.25 Using Processing State Transition Diagrams", Australian Telecommunication Research, Vol. 14, No. 2, 1980, or in [SUN-81], pp. 421-437.

[ECS-TAR-86] Ecsedi Tóth P., Tarnay K.: "Relations between Formal Description of Protocols and Performance Evaluation", Seminar on Modeling and Performance Evaluation of Distributed Computer Systems, Wendisch Rietz, 1986.

[ECS-TAR-87] Ecsedi Tóth P., Tarnay K.: "Formal Description Oriented Performance Evaluation of Protocols", Report KFKI-1987-46, Budapest, 1987.

[EHR-86] Ehrig H., Mahr B.: "Fundamentals on Algebraic Specification 1". Springer Verlag EATCS 6, Berlin, 1985.

[FAV-LINN-87] Favreau J-P., Linn R.J.: "Automatic Generation of Test Scenario Skeletons from Protocol Specifications Written in Estelle", in [SAR-BOCH-87], pp. 191-202.

[ÇAR-75] Gardner R.I.: "A Methodology for Digital System Design Based on Structural and Functional Modeling", Ph. D, Thesis, UCLA-ENG-7488, Jan. 1975.

[GIE-83] Giessler A.: "Testing and Diagnosis Aids for Higher Protocols", in [RUD-WE-83], pp. 407-420.

[GILL-62] Gill A.: Introduction to the Theory of Finite State Machines. McGraw Hill, New York, 1962.

[GILL-63] Gill A.: "Time Varying Sequential Machines", Journal of the Franklin Institute, No. 276, 1963, pp. 516-539.

[GIN-66] Ginsburg S.: The Mathematical Theory of Context-free Languages, McGraw Hill, New York, 1966.

[GOS-71] Gostelow K.P.: "Flow of Control, Resource Allocation and the Proper Termination of Programs", Ph. D. Thesis, UCLA - ENG - 7179, 1971.

[GOTZ-85] Gotzhein R.: "Modellierung und Spezifikation von Diensten und Verhalten in verteilten Systemen", Ph. D. Thesis, 1985.

[GOTZ-87] Gotzhein R.: "Specifying Abstract Data Types with LOTOS", in [SAR-BOCH-87], pp. 15-26.

[GÖ-70] Gönenc G.: "A Method for the Design of Fault Detection Experiments", IEEE Trans. on Computers, June 1970, pp. 551-558.

[GUT-77] Guttag J.V.: "Abstract Data Types and the Development of Data Structures", Comm. of the ACM, Vol. 20, No. 6, 1977, pp. 396-404.

[HAR-76] Harangozó J.: "The Description of Data Link Protocols for Computer Networks Using Formal Languages", Alkalmazott Matematikai Lapok, 1976, pp. 315-329 (in Hungarian).

[HAR-77] Harangozó J.: "An Approach to Describing a Data Link Level Protocol with a Formal Language", Fifth Data Comm. Symp., Snowbird, Utah, 1977.

[HAR-78a] Harangozó J.: "Protocol Definition with Formal Grammars.", Computer Network Protocols, A. Danthine (ed.), Liege, 1978.

[HAR-78b] Harangozó J.: "The Formal Description of Protocols for Computer Networks", Ph. D. Thesis, Tech. Univ. of Budapest, Budapest, 1978 (in Hungarian).

[HOF-70] Hoffmann H.J.: "On Linguistic Aspects of Communication Line Control Procedures",IBM-Report, RZ 345, 1970.

[HOL-CO-70] Holt A., Commoner F.: "Events and Conditions", Report of Applied Data Research, New York, 1970.

[JAR-86] des Jardins R.: "Towards the Information Society: World Cooperation on Open Systems Standardization", in [CSA-86], pp. 15-18.

[KEL-76] Keller R.M.: "Formal Verification of Parallel Programs", CACM, 7, 1976, pp. 371-384.

[KOE-TAR-87] Koenig H., Tarnay K.: "An intelligent protocol workstation", Report KFKI-1987-45, Budapest, 1987.

[KOH-78] Kohavi Z.: Switching and Finite Automata Theory. McGraw Hill, New York 1978.

[KOV-87] Kovács I.: "Description of Transport Protocol Class 0 in SDL", Technical Report, KFKI, Budapest 1987.

[KOV-82] Kovács L.: "The Formal Specification and Verification of Computer Network Protocols", M. Sc. Thesis, Budapest, 1982 (in Hungarian).

[LEMO-73] Le Moli G.: "A Theory of Colloquies", First European Workshop on Computer Networks, Arles, 1973.

[LI-CO-83] Linn R.J. McCoy W.H.: "Producing tests for implementations of OSI protocols", in [RUD-WE-83], pp. 505-520.

[LI-NI-83] Linn R.J., Nightingale I.S.: "Some Experience with Testing Tools for OSI Protocol Implementations", in [RUD-WE-83], pp. 521-531.

[LINN-85] Linn R.J.: "An Evaluation of the ICST Test Architecture after Testing Class 4 Transport", in [YEM-85], pp. 611-622.

[LINN-86] Linn R.J.: "The Features and Facilities of ESTELLE", in [DIAZ-86], pp. 271-298.

[LYNCH-68] Lynch W.C.: "Reliable Full-Duplex File Transmission over Half-Duplex Telephone Lines", Commun. ACM, Vol. 11, No. 6, 1968, pp. 407-410.

[LYNCH-69] Lynch W.C.: "Commentary on the Foregoing Note", Commun, ACM, Vol. 12, No, 5, 1969. p. 261.

[LYO-82] Lyons J.: Chomsky. Collins & Sons, Glasgow, 1982.

[MAR-8] Margitics I.: "The Analysis of an End-to-end Transport Protocol for Heterogeneous Computer Networks", M. Sc. Thesis, Budapest, 1982 (in Hungarian).

[MEL-PET-84] Melendez W.A., Peterson E.L.: "Tutorial on the Upper Layers of the OSI Reference Model", CEC Report, 1984.

[MEER-86] de Meer J.: "Tutorial on LOTOS data types" Draft, 1986.

[MER-74] Merlin P.M.: "A Study of the Recoverability of Computing Systems", Ph. D. Thesis, Univ. of California, Irvine, 1974.

[MER-76] Merlin P.M.: A Methodology for the Design and Implementation of Communication Protocols. IEEE Trans. on Communications, Vol. COM-24, No. 6, 1976.

[MER-79] Merlin P.M.: "Specification and Validation of Protocols", IEEE Trans. on Communications, Vol. COM-27, No. 11, Nov, 1979, pp. 1671-1680.

[MER-FAR-76] Merlin P.M., Farber D.J.: "Recoverability of Communication Protocols – Implication of a Theoretical Study", IEEE Trans. on Communications, Vol. COM-24, September 1976, pp. 1036-1043, or in [SUN-81], pp, 127-141.

[MEZ-SCHRE-73] Mezzalira L., Schreiber F.A.: "Designing Colloquies", First European Workshop on Computer Networks, Arles, 1973.

[MILN-80] Milner R.: "A Calculus of Communicating Systems", Lecture Notes in Computer Science, Vol. 92, Springer Verlag, New York, 1980.

[NAGY-76] Nagy Á.: "The Simulation of Complicated Systems with Modified Petri Nets.", M. Sc. Thesis, Budapest, 1976 (in Hungarian).

[NAI-81] Naito S., Tsunoyama M.: "Fault Detection for Sequential Machines by Transition Tours", IEEE 1981, pp. 128-143.

[NI-82] Nightingale J.S.: "Protocol Testing Using a Reference Implementation", in [SUN-82], pp. 513-522.

[NUTT-72] Nutt G.J.: "Evaluation Nets for Computer System Performance Analysis", in AFIPS Conf. Proc. Vol. 41, part 1, 1972, pp, 279-286.

[PAR-GU-85] Parrow J., Gustavsson R.: "Modeling Distributed Systems in an Extension of CCS with Infinite Experiments and Temporal Logic", in [YEM-85], pp. 309-348.

[PAY-LI-78] Payne A.D., Lings B.J.: "Topological Design of a Small Resource Sharing Network" Australian Computer Journal, Vol. 10, No. 4, Nov. 1978.

[PET-77] Peterson J.L.: "Petri Nets", Ass. Comput. Mach. Comouting Surveys, Vol. 9, Sept. 1977. pp. 223-251.

[PET-81] Peterson J.L.: "Petri Net Theory and the Modeling of Systems", Prentice-Hall, Inc., Englewood Cliffs, N.J. 1981.

[PETRI-62] Petri C.: "Kommunikation mit Automaten", Ph. D, Thesis, University of Bonn, 1962.

[POS-74] Postel J.B.: "A Graph Model Analysis of Computer Communications Protocols", Ph. D. Thesis, UCLA-ENC-7410, 1974.

[PU-PO-80] Puzman J., Porizek R.: Communication Control in Computer Networks. John Wiley & Sons, New York, 1980.

[RAY-82] Rayner D., "A System for Testing Protocol Implementation", Computer Networks 6, Dec. 1982.

[RAY-83] Rayner D.: "Towards an Objective Understanding of Conformance", in [RUD-83], pp. 477-492.

[RAY-86] Rayner D.: "Standardizing Conformance Testing for OSI", in [CSA-86], pp. 47-68.

[RAY-HA-81] Protocol Testing - Towards Proof? Rayner D., Hale R.W.S. (eds.) Report NPL, Teddington, 1981.

[RAZ-ES-80] Razouk R., Estrin G.: "Validation of the X.21 Interface Specification Using SARA", NBS Trends and Applications Symposium, May 1980; or in [SUN-81], pp. 243-251.

[RAZ-ES-82] Razouk R.R., Estrin G.: "Modeling and Evaluation of Communication Protocols", in [SCHO-82] pp. 167-190.

[REV-79] Révész Gy.: An Introduction to the Theory of Formal Languages. Akadémiai Kiadó, Budapest, 1979 (in Hungarian).

[RUD-85] Rudin H.: "An Improved Algorithm for Estimating Protocol Performance", in [YEM-85], pp. 515-526.

[RUD-WE-83] Protocol Specification, Testing, and Verification, III. Rudin H., West C.H. (eds.) North-Holland Publ. Co., Amsterdam, 1983.

[RUD-WE 87] Protocol Specification, Testing, and Verification VII. Rudin H., West C.H. (eds.) North-Holland Publ. Co., Amsterdam, 1987.

[RUS-LA-74] Rusbridge R.E., Langsford A.: "Formal Representation of Protocols for Computer Networks", AERE-R 7826, 1974.

[SARA-84] Saracco R.: "Course on SDL, Specification and Description Language", CSELT Report, Turin, 1984.

[SAR-84] Sarikaya B.: "Test Design for Computer Network Protocols", Ph. D. Thesis, McGill Univ., Montreal, 1984.

[SAR-BOCH-82] Sarikaya B., Bochmann G.V.: "Some Experience with Test Sequence Generation for Protocols" in [SUN-82], pp. 555-567.

[SAR-BOCH-87] Protocol Specification, Testing, and Verification, VI. Sarikaya B., Bochmann G.V. (eds.) North-Holland Publ, Co., Amsterdam, 1987.

[SCHO-82] Computer Networks and Simulation II. S. Schoemaker (ed.) North-Holland Publ. Co., Amsterdam, 1982.

[SCO-PAP-86] Scollo G., Pappalardo G., Logrippo L., Brinksma E.: "The OSI Transport Service and its Formal Description in LOTOS", in [CSA-86], pp. 465-488.

[SCO-SIN-87] Scollo G., Van Sinderen M.: "On the Architectural Design of the Formal Specification of the Session Standards in LOTOS", in [SAR-BOCH-87], pp. 3-14.

[SEB-87] Sebestyén B.: Local Area Networks. Műszaki Könyvkiadó, Budapest, 1987 (in Hungarian).

[SHER-85a] Sherif M.H., Miller V., Woo E.: "X.25 Protocol Verification Facility for AT&T No.1 PSS", in [YEM-85], pp. 69-77.

[SHER-85b] Sherif M.H., Hoover G.L., Wiederhold R.P.: "X.25 Conformance Testing - a Tutorial", IEEE Comm. Mag. Vol. 24, No. 1, 1985, pp. 16-27.

[SID-BLU-86] Sidhu D.P., Blumer T.V.: "Verification of NBS Class 4 Transport Protocol", IEEE Trans on Comm., Vol. COM-34, No. 8, Aug. 1986, pp. 781-789.

[STA-85] Stallings W.: Data and computer communications. Macmillan Publ. Co., New York, 1985.

[STE-86] Steinacker M.: "SDL and ESTELLE: State Transition Oriented Specification Techniques in CCITT and ISO", in [CSAB-86], pp. 489-505.

[SUN-81] "Communication Protocol Modeling" Sunshine C. (ed.). Artech House, Washington, 1981.

[SUN-82] "Protocol Specification, Testing, and Verification" Sunshine C. (ed.) North-Holland Publ. Co., Amsterdam, 1982.

[SYM-80] Symons F.J.W.: "The Verification of Communication Protocols Using Numerical Petri Nets", Australian Telecommunication Research, Vol. 14, No.1, 1980. pp. 34-38, or in [SUN-81], 149-153.

[SYM-81] Symons F.J.W.: "Introduction to Numerical Petri Nets, a General Graphical Model of Concurrent Processing Systems", Australian Telecommunication Research, Vol. 14, No. 1, 1980. or in [SUN-81], pp. 143-147.

[TAR-76] Tarnay K.: "Computer Networks", Techn.Report KFKI-76-51, Budapest, 1976.

[TAR-83] Tarnay K. et al.: "New Results on Computer Communication", Report KFKI-1983-26, Budapest, 1983.

[TAR-MON-85] Tarnay K., Mondvai G.: "Protocol Testing", Technical Memorandum of SWIFT-II, Brussels, 1985.

[TAR-87a] Tarnay K., Kovács I., Wagner-Dibuz S., Wittmann R.: "Conformance Testing", Technical Report, Budapest, 1987 (in Hungarian).

[TAR-87b] Tarnay K., Wittmann R., Kovács I., Wagner-Dibuz S.: "Conformance test centers", Technical Report, Budapest, 1987 (in Hungarian).

[TAR-87c] Tarnay K., Kovács I., Kicsiny-Törő M.: "Implementation and testing of X.25 Protocols", Technical Report, Budapest, 1987 (in Hungarian).

[TAR-87d] Tarnay K. et al.: "Network Environment Manipulator", Report KFKI-1987-13, Budapest, 1987 (in Hungarian).

[TAR-ECS-87] Tarnay K., Ecsedi Tóth P.: "Observer Specification in LOTOS", Invited paper of PD'87 ELTE, Budapest, 1987.

[TAR-WAG-87] Tarnay K., Wagner-Dibuz S.: "An SDL-based Protocol Tester", Proceedings of the SDL Forum, Hague, 1987.

[UR-PRO-84] Ural H., Probert R.L.: "Testing Specifications and Designs of Communication Protocols and Services", Report, University of Ottawa, 1984.

[VEN-PI-86] Venkatraman R.C., Piatkowski T.F.: "A Formal Comparison of Formal Protocol Specification Techniques", in [DIAZ-86], pp. 401-420.

[VI-TE-BO-83] Vissers Ch., Tenney R.L., Bochmann G.v.: "Formal Description Techniques", Proc. of the IEEE, Vol. 71, No. 12, 1983, pp. 1356-1364.

[WAG-87] Wagner-Dibuz S.: "Specification and Testing of Transport Protocols", Report KFKI-1987-20, Budapest, 1987 (in Hungarian).

[YE-ZA-78] West C.H., Zafiropulo P.: "Automated Validation of a Communications Protocol: The CCITT.21 Recommendations", IBM Journal of Research and Development, Vol. 2, January 1978, pp. 60-71.

[WHE-CHI-84] Wheeler G., Chin I.P.W., English K.: "Specification of the Link Access Procedures of X.25 Using SDL", Telecom Australia - Report 7648, 1984.

[WHE-SYM-82] Wheeler G., Symons F.J.W., Billington J.: "Description of Numerical Petri Nets", CCITT Question 39/VII. Dec. 1982. (Delayed Contribution No. 363).

[YEM-85] "Protocol Specification, Testing, and Verification", IV. Yemini Y., Strom R., Yemini S. (ed.), North-Holland Publ. Co., Amsterdam, 1985.

[ZIM-80] Zimmermann H.: "OSI Reference Model - The ISO Model of Architecture for Open Systems Interconnection", IEEE Trans. on Comm. Vol. COM-28, April 1980. pp. 425-432.

STANDARDS

ISO, Information Processing Systems - Open Systems Interconnection - Basic Reference Model
IS 7498 (1984)

ISO, Information Processing Systems - Open Systems Interconnection - ESTELLE, a formal description technique based on an extended state transition model
2nd DP 9074 (1986)

ISO, Information Processing Systems - Open System Interconnection - LOTOS - A formal description technique based on temporal ordering of observational behaviour
DP 8807 (1986)

ISO, OSI Conformance Testing Methodology and Framework
2nd DP 9646 ISO TC 97/SC 21 N, Oct. 1987

ISO, Information Processing Systems - Open Systems Interconnection - Connection Oriented Transport Protocol Specification
DP 8073, ISO/TC 97/SC 16 N 1433, July 1983

ISO, Information Processing Systems - Open Systems Interconnection - Specification of Abstract Syntax Notation One
DIS 8824.2, July 1986

ISO, Information Processing Systems - Open Systems Interconnection - File Transfer, Access and Management
DIS 8571, August 1986.

ISO, Data Communication - High level data link control unbalanced classes of procedures
DP 6159, 1980

ISO, Data Communication - HDLC balanced classes of procedures
DP 6256, 1980

CCITT V.24.
List of definitions for interchange circuits between data-terminal equipment and data-circuit terminating equipment.

CCITT X.21.
Interface between data-terminal equipment and data-circuit terminating equipment for synchronous operation on public data network.

CCITT X.25.
Interface between data-terminal equipment and data-circuit terminating equipment for terminals operating in the packet mode on public data networks.

CCITT X.75.
Terminal and transit call control procedures and data transfer system on international circuits between packet switched data networks.

CCITT X.200
Open System Interconnection for CCITT Applications - Reference Model

CCITT SG XI, Z.100-Z.104
SDL - Specification and Description Language - Basic Concepts - Structuring Concepts - Refinement - Data

General Information, Binary Synchronous Communication (BSC), IBM Systems Reference Library, From No. A27-3004.

ABBREVIATIONS

ABM	Asynchronous Balanced Mode
ACK	Acknowledgement
ACK-WAIT	Acknowledgement-wait state
ACT. IND	Activate Indication
ACT. REQ	Activate Request
ADI	Agence de l'Informatique
ADT	Abstract Data Types
AIR RES	Air Reservation
ANSI	American National Standards Institute
ARM	Asynchronous Response Mode
ASCII	American Standard Code for Information Interchange
ASN. 1	Abstract Syntax Notation One
ASP	Abstract Service Primitive
BCC	Block Check Characters
BER	Bit Error Rate
BLER	Block Error Rate
BSC	Binary Synchronous Communications
BTAM	Basic Telecommunication Access Method
CASE	Common Application Service Elements
CC	Connection confirmation
CCITT	International Telegraph and Telephone Consultative Committee
CCS	Calculus of Communicating Systems
CEP	Connection End Point
CHILL	CCITT High Level Language
CMDR	Command Reject
CON conf	Connection confirmation
CON req	Connection request
CR	Connection Request
CRA	T-CONNECT request from user A
CRC	Cyclic Redundancy Checking
CT	Communication Tree
D	Delivery Confirmation Bit (X.25 Packet level)
DCE	Data-Circuit Terminating Equipment
DEACT. IND	Deactivate Indication
DEC	Digital Equipment Corporation
DFG	Data Flow Graph

DFL	Data Field Length
DIA	T-DISCONNECT indication sent to user A
DISC	Disconnection
DISC ind	Disconnection indication
DISC req	Disconnection request
DLE	Data Link Escape
DM	Disconnected Mode
DR	Disconnect Request
DT	Data (Transport PDU)
DTE	Data-Terminal Equipment
DT ind	Data indication
DT req	Data request
EAB	Empty Queue A to B
EBCDIC	Extended Binary Coded Decimal Interchange Code
EMU	Environment Manipulator Unit
EOT	End of Text
ERR	Error
ESTELLE	Extended Finite State Machine Specification Language
ETB	End of Text Block
ETX	End of Text
FCS	Frame Check Sequence
FDT	Formal Description Techniques
FDX	Full-duplex
FIFO	First in first out
FLC	Frame Level Control
FRMR	Frame Reject
FSM	Finite-State Machine
FTAM	File Transfer Access and Management
GMB	Graph Model of Behaviour
GMD	Gesellschaft für Mathematik und Datenverarbeitung
HDX	Half-duplex
HDLC	High-Level Data Link Control
I	Information frame (HDLC)
IBM	International Business Machines Corporation
ICI	Interface Control Information
ID	Interface Data
IDU	Interface Data Unit
IS	Initial Sequence
ISO	International Standards Organization
ITB	End of Intermediate Transmission Block
IUT	Implementation Under Test

JM	Job Manipulation
JTM	Job Transfer and Manipulation
LAPB	Link Access Procedure Balanced
LOTOS	Language for Temporal Ordering Specification
M	More Data Bit (X.25 Packet level)
MML	Man-Machine Language
NAK	Negative Acknowledge
NBS	National Bureau of Standards
NDRA	NDATA flow control ready for A
N-D req	Network Data request
NISL	Network Interface Sub-Layer
NPL	National Physical Laboratory
NPN	Numerical Petri Net
N(R)	Receive number (HDLC)
NRCC	Network Routing Control Centre
NRM	Normal Response Mode
N(S)	Send number (HDLC)
NSAP	Network Service Access Point
NSDU	Network Service Data Unit
OSI	Open Systems Interconnection
OSIA	Open Systems Interconnection Architecture
PCI	Protocol Control Information
PDU	Protocol Data Unit
P/F	Poll/Final bit
PID	Process Instance Identifier
P(R)	Receive number (X.25 Packet level)
PRED	Predecessor
P(S)	Send number (X.25 Packet level)
PSAP	Presentation Service Access Point
PT of Sale	Point of Sale
RC	Retransmission number
REJ	Reject
RM	Reference Model
RNR	Receive Not Ready
RR	Receive Ready
RST	Rigid Synchronization Tree
RX-one	Reception of bit one
RX-zero	Reception of bit zero

SABM	Set Asynchronous Balanced Mode
SABME	Set Asynchronous Balanced Mode Extended
SAP	Service Access Point
SARM	Set Asynchronous Response Mode
SASE	Specific Application Service Elements
SDL	Specification and Description Language
SDL/GR	SDL Graphic Syntax
SDL/PE	SDL Pictorial Element Syntax
SDL/PR	SDL Program-like Syntax
SDU	Service Data Unit
SNA	Systems Network Architecture
SNRM	Set Normal Response Mode
SREJ	Selective Reject
SSAP	Session Service Access Point
ST	Synchronization Tree
SUCC	Successor
SXRM	SNRM or SARM
TC	Transport Connection
T-C conf	Transport Connection Confirmation
T-C ind	Transport Connection Indication
T-C req	Transport Connection Request
T-C resp	Transport Connection Response
TESDI	Testing and Diagnosis Aid
TPDU	Transport Protocol Data Unit
TS	Transport Service
TSAP	Transport Service Access Point
UA	Unnumbered Acknowledge
UD	User Data
USSE	User Specific Service Elements
VF	Virtual File
VT	Virtual Terminal
VTS	Virtual Terminal Service

INDEX

abstract machine 340-342
abstract representation 72
acknowledgement 22, 36, 74, 86, 99, 247, 268
ACT ONE 160-168, 213
    ADTs 162-163
    atomic specification 163-166
    complex specification 166-168
active tester 98
addressing 16-17
a-sequence 327-330
alternating bit protocol 229-246
application layer 30, 70-76
application process 71
attribute vector 139
automatic testing 306, 312
axioms 187-188

basic set theory 108-109
behaviour expression 153-157, 217-222
behaviour tree 152-153
b-sequence 327, 330-331
bilateral agreement 54
bit error rate 80, 86
bit-oriented protocol 37-38
block error rate 80
blocking 22

CCS 151-159, 213
    equivalences 157-158
characteristic function 118, 282
character-number-oriented protocol 37
character-oriented protocol 37
Chomsky classes of grammars 170-173
cohesion 82-83
complementary test sequence 321
computer network 3-11
    classification 3-4
    operation 7-10
    origin 4-7

concrete representation 72
conformance tester 97-99, 321
conformance testing 78, 89-96
abstract model 93-96
classification 91-92
requirements 89-91
connection establishment and release 18, 37, 48, 54, 59, 67, 247
connection establishment delay 101
connectivity 82, 103
context-free grammar 170
context-sensitive grammar 170
control flow 117, 293
control flow graph 142

data flow graph 142-146, 293-294
data flow program 145
data generator 308
datagram 54
data line monitor 305
data link layer 29, 36-52
data transfer 33-34, 92
data units 19-21
deadlock 125
degree of node 82
disconnection delay 102
distinguishing sequence 115
distributed system 32
distributed test method 95-96

enabling conditions 141, 263, 292-293
enabling predicate 237
E-net 136-140
error detection 36, 37, 43, 59
error generator 82, 87-89
error handling 22, 25, 92, 99
ESTELLE 151, 200-212
channel 206
event gates 204-206
examples 208-211, 237-246
modules 204, 241-243
software tools 212
specification 206-208
transitions 202-204, 243-246
exception generator 308
expedited data transfer 59, 63, 66
extended mode 46
extended state-transition graph 285-286

finite-state machine 107-109, 110, 114, 281-284, 322
equivalence 114-116
properties 114
firing 123-125, 139, 293
fixed routing 56
flooding 56
flow control 10, 21, 25
formal language 169-173
classification 170-173
grammar 169-171

Gönenc method 326-339
grammatical rule system 169

HDLC models 247-280
formal grammar 252-263
heterogeneous systems 10, 68
homogenous system 10
hybrid 247-252
hybrid model 247-252
SDL/GR 263-280

incidence matrix 130
inheritance in SDL 188-191
initial sequence 314, 327
initial testing 305-306
interaction 152, 213, 216
interaction point 197, 206, 214, 301
interface tester 305

layer elements 15-16
functions 18-22
services 22-24
link-set-up module 248-250
load capacity 80
local test method 93-94
LOTOS 213-225
behaviour expressions 217-222
data types 222-224, 297-300
interactions 213-217, 301
processes 213-215, 301-303
recursion, nondeterminism 224-225
synchronization 213-215

management 13, 67
manual conformance tester 310-311, 312

marking 123, 125
modified Petri nets 132-141
module 197, 204-205
multiplexing 19, 53, 59

NBS test system 306-309
network environment 79-82
network layer 29, 53-58
normal data transfer delay 102
NRCC 56-57
numerical Petri net 140-141, 288-293

operators in SDL 186-188

packet formats 54-56
partitioning 180-182
passive testers 305
performance measurement 79, 81, 99
    parameters 101-103
    requirements 99-100
    scheduling 100-101
Petri net 123-131
    examples 146-148, 234-240
    features 125-127
    mathematical description 129-131
    subgraphs 127-128
physical layer 29, 32-35
Poll/final bit 45-47
predefined data types 185-186
presentation layer 29, 68-70
process 151, 213-214, 264-266
production rules in table form 257-259
protocol 15, 25-28
    features 26-27
    parameters 27
protocol data units 19-21, 93
protocol handler 316
protocol tester 87-89, 305-306

queueing system 80-81

Reference Model 11-12, 32, 86, 97, 197
reference protocol 87, 93, 306
refinement in SDL 182-183
regular grammar 171
remote test method 93, 95-96

"rendezvous" principle 152
report generator 316
resetting 21, 53, 99
robustness 79
routing 56-57

safe marking 125
scenario file 308
scenario interpreter 309
SDL 175-196, 286-287
basic concepts 176-177
concrete syntax 176
data concepts 184-186, 269-277
data hierarchy 191
general definition 175-176
process diagram 178-179, 277-279, 287
structuring concepts 180-183
tools 192-196
segmenting 21, 53, 60
sequence numbering 36-37, 45, 53, 229-230, 254-257
service primitive 22-23, 93
session layer 29, 66-68
sort 161-162, 184
splitting 19, 37
state-transition graphs 111-114, 230, 284
state-transition tables 111-114, 233, 283
state-transition table controlled conformance tester 311
stimulus 324
stochastic routing 56
strongly connected graph 282
successor tree 116-121
suffix 17

test centre 310, 313-319
test centre driver 308
test driver 87, 97, 312
test logger 316
test responder 87, 97, 312
test scenario 308-309
test sequence 99, 318-319, 321-342
test sequence generation 321-342
throughput 102
time Petri net 135-136
time sequence diagram 24
title 16-17
topological matrices 79

traffic measurement 82
transfer syntax 68
transition 123-124, 128, 136-138, 292
transition tour method 321-324
transparent transmission 43-53
transport layer 29, 59-66
transport protocol 59-66, 281-303
trap 126
Turing machine 109-110

virtual call 53
virtual circuit 53

window mechanism 37
W-method 324-325

MIX
Papier aus verantwortungsvollen Quellen
Paper from responsible sources
FSC® C105338

If you have any concerns about our products,
you can contact us on
ProductSafety@springernature.com

In case Publisher is established outside the EU,
the EU authorized representative is:
Springer Nature Customer Service Center GmbH
Europaplatz 3, 69115 Heidelberg, Germany

Printed by Libri Plureos GmbH
in Hamburg, Germany